HOME & DRY

Birgit Bulla is a journalist who lives in Munich and works as a freelance editor for various magazines. Out of the blue in her mid-twenties, she developed an irritable bladder. The response to her blog, pinkelbelle.de, shows that she's far from alone. Today she knows everything about this part of the body, and it comes as no surprise: Bulla is, after all, the Latin word for bladder.

Rachel Stanyon has worked as a teacher and researcher in Germany and the UK, and is currently based in Australia, where she volunteers for the world literature journal *Asymptote* and works as a translator from German. She holds a master's in translation, and in 2016 won a place in the New Books in German Emerging Translators Programme.

HOME & DRY

Heal Your Bladder, Treat UTIs and Incontinence, and Improve Your Health

BIRGIT BULLA

translated by Rachel Stanyon

with illustrations by Annette Bulla

SCRIBE
Melbourne • London

Scribe Publications
2 John St, Clerkenwell, London, WC1N 2ES, United Kingdom
18–20 Edward St, Brunswick, Victoria 3056, Australia
3754 Pleasant Ave, Suite 100, Minneapolis, Minnesota 55409, USA

First published by hanserblau in Carl Hanser Verlag as
Noch ganz dicht? Alles Wissenswerte über die Blase in 2020
First published in English by Scribe 2022

Copyright © hanserblau in der Carl Hanser Verlag, Munich 2020
Internal illustration copyright © Annette Bulla 2022
Translation copyright © Rachel Stanyon 2022

The advice in this book is not intended to replace the services of trained
health professionals or be a substitute for medical advice.
You are advised to consult with your healthcare professional with regard
to matters relating to your health, and in particular regarding matters that
may require diagnosis or medical attention.

Typeset in Minion Pro by the publishers

Printed and bound in the UK by CPI Group (UK) Ltd, Croydon CR0 4YY

Scribe is committed to the sustainable use of natural resources and
the use of paper products made responsibly from those resources.

978 1 950354 93 1 (US edition)
978 1 913348 76 2 (UK edition)
978 1 922310 81 1 (AU edition)
978 1 922586 47 6 (ebook)

Catalogue records for this book are available from the
National Library of Australia and the British Library.

scribepublications.com
scribepublications.co.uk
scribepublications.com.au

CONTENTS

To Anyone with a Problem Bladder: You Are Not Alone!

I've been needing to go non-stop since I turned twenty-seven. I'm one of those people who always wants the aisle seat at the cinema or on a plane. I'm the person who dashes to the toilet for just one more quick pee before she leaves the house. The person with the irritable bladder. What does that mean? My bladder rules my life. It pressures me into going to the toilet every half hour — and super urgently, too. It revs from zero to one hundred in seconds, like a Ferrari — but one I can't brag about.

Sadly, it's not that uncommon for bladders to go haywire. Urinary tract infections are the second most common illness to prompt women to visit a doctor, and experts now consider incontinence a common condition. But issues related to the bladder still don't get much attention, and many women and girls find them difficult to talk about. Why? Because of a lack of education.

In the meantime, though, a lot has been going on downstairs: books, articles and blogs have been getting to the bottom of our vaginas, explaining their qualities and quirks.

1

Our periods have become an acceptable symbol of woman-hood, which we celebrate with stylish and beautiful products. Hiding your loud, pink box of tampons in the bathroom before visitors arrive? So 2012. Accepting and celebrating the female body as it is — hairs, dimples, discharge and all — has never been as hot as it is now. Of course, this is fantastic for us and our vaginas. But what about our bladders? They continue to eke out a shy, anonymous existence, looking on enviously at the success of their colleagues.

In medical literature, the urinary tract is usually discussed from a male perspective. If you flick through urology textbooks, you can learn in great detail about why men have trouble going to the toilet. Spoiler alert: it often has something to do with the prostate. Which is of course super important, but only for men.

Why is it like this? Probably because medical research was designed around the male body from the very beginning. Did you know that scientists used to think women were just a smaller, more fragile version of men? In reality, there are physiological differences between the male and female body, and women often display completely different symptoms. This is especially true for the bladder. Even pop-science medical books — you know, the ones mere mortals can understand — are pretty bad when it comes to the female bladder. Wherever you look, you usually either come up against the prostate and the male urinary system again, or find books that instead focus on sex …

So, it's high time we paid the female bladder the attention she deserves. Did you know that it can be dangerous not to go to the toilet often enough? Or that infrequent urination might be telling you something about your psyche? What our bladder accomplishes in a day really is astounding.

I had never before looked at the human body as intensively as I did while researching this book. I pulled all-nighters on Internet forums. I spoke with all kinds of experts and pored over books that are usually only used by students swotting for their exams. These days, I almost think it's a pity that I was too lazy to study medicine. (Ok, let's be honest: my marks weren't good enough, either.) Perhaps I would have made a great urologist. But life as an editor — which is my actual job, when I'm not reading everything I can about the bladder — is really pretty good, too.

A word of caution, though: this is not a medical textbook, and I'm not a doctor, I'm a patient. I'm sharing my own suffering in this book, and I hope it'll be like the friend that you'd most like to have in the waiting room with you, the one who knows what you're going through and whom you can ask any question. What's life like when you perpetually need to pee? What kind of doctor should you go to first, and what will happen there? What happens in examinations, which treatments are available and what do they feel like? Aside from this, I also want to point you in the direction of some possible solutions that doctors don't always offer. You might be able to contribute to your own health by thinking outside the box (or the bladder) and staying persistent.

Meanwhile, I've had to go to the toilet three times since I started writing this …

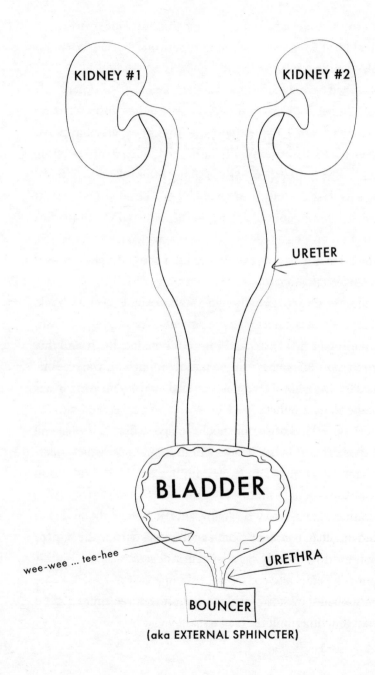

1.
The Bladder and Its Co-Workers

The urinary tract is one of the sexiest organs we humans have. In terms of its anatomical structure, it basically looks like a sculpture, or a trophy you'd be proud to put on your shelf. At the top sit two perfectly mirrored oval kidneys, each of which is connected by a delicate ureter to the bladder below. All this is rounded off by the funnel-shaped urethra, which, punctuated by the urethral sphincters, curves like a swan's neck down to the bladder outlet.

Its charm and good looks haven't gone to the urinary tract's head, though — it doesn't think anything is beneath it. Its main areas of responsibility? Waste water treatment, temporary storage, protection and dispatch. It doesn't receive much attention for this, though — as long as it does its work diligently and conscientiously, that is. It's only when they stop working properly that we start thinking about the bladder, kidneys and co.

A NOTE ON UTI TERMINOLOGY

As you can see, the urinary tract includes the kidneys, two ureters, a bladder, and a urethra. **Urinary tract infections (UTIs)** can affect any part of the system, but most commonly impact the lower urinary tract — the bladder and the urethra. **Cystitis** is another name for inflammation of the bladder. It is usually caused by a bladder infection, and is a common kind of UTI. We'll learn all about cystitis and its nasty friends in Chapter 3.

The Kidneys:
The Body's Waste Water Treatment Plant

The kidneys' chief task is to filter waste products from the blood and turn them into urine. Renal corpuscles are responsible for this: they sit on the outer renal cortex and spend the whole day scanning our blood for harmful substances and filtering them out. It's a truly mammoth task: each day, all our blood — and we're talking five to six litres of the stuff — flows through the kidneys about 300 times. In total, those little renal corpuscles filter roughly 1,700 litres of blood every day. And you thought you had a stressful job!

The contaminants filtered out through this process are turned into primary urine, a precursor of secondary urine, *i.e.* the finished urine that we later pee away. This primary urine passes through what is known as the renal tubule. This is where the brewing process itself takes place, through which

150 to 180 litres of primary urine are produced each day. From this primary urine, everything that the body thinks it might be able to use after all is fished back out. Approximately 99 per cent of this filtrate is recycled back into the body — most of the material reclaimed is water, but it also includes molecules such as sugars, small proteins, and minerals like sodium and magnesium. If we didn't get the water back again, it wouldn't be long before we died.

Once this process has been completed successfully, the secondary urine — 0.5 to 2 litres per day — is transported through the collecting duct system. On the way, more water is extracted from it, until it is collected in the renal pelvis in an even more concentrated form. From here, the ureters, one of which is attached to each kidney via a little indentation, pump it along to the bladder.

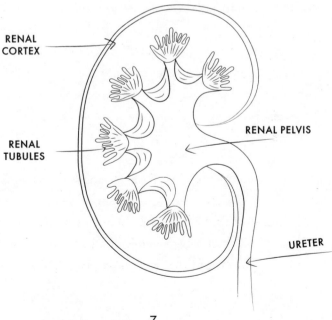

RENAL CORTEX

RENAL PELVIS

RENAL TUBULES

URETER

As if filtering, recycling, and brewing urine weren't already enough, like any true workaholic, the kidneys also have other important jobs, such as regulating our blood pressure by filtering more or less water out of the blood. This is how it works: when our blood vessels retain more water, the volume in them increases and our blood pressure goes up. When the kidneys siphon off more water, the volume in our blood vessels decreases and our blood pressure drops. In addition to this, the kidneys also produce the hormones erythropoietin, which encourages the formation of red blood cells in our bone marrow; and calcitriol, an active form of vitamin D which regulates the amount of calcium in our body, making it super important for bone development. And, on top of all that, our kidneys help maintain what's known as the acid–base balance (roughly speaking, our pH levels) during the filtration process by ensuring that our blood is neither too acidic nor too alkaline. They're pretty great, those kidneys!

The Ureters:
The Bladder's Pipes

The ureters are the connecting link between the kidneys and the bladder, and pump the finished urine out of the renal pelvis. Each of the two ureters is between 25 and 30 centimetres long and has a diameter of two to four millimetres. They always remind me a bit of a flower stem. So, if the bladder is the roots, the kidneys are the flowers … but whatever, back to the ureters!

To keep the urine flowing in the direction of the bladder and not back up the other way, the ureters are fitted out with smooth muscle that contracts in waves, ensuring the urine is safely transported into the bladder without taking any detours. This so-called peristaltic wave runs through the ureters several times each minute, so the tubes are always in motion.

At the entrance to the bladder, the ureters are woven into the bladder muscle to form a kind of valve that prevents the urine already inside from getting back into the ureters. In case you've ever wondered if you could pee while doing a headstand: these valves make it possible.

The Bladder:
Muscular Tupperware for Our Urine

The bladder is a hollow organ situated in the middle of the lower abdomen behind the pubic bone, resting on the pelvic floor. The bladder's main job is to store urine, wait until enough has been collected, and then discharge it at the right moment. The technical term for this is 'micturition'.

To help you pee hassle-free, the bladder has a whole lot of tricks up its sleeve. It's coated with a soft layer of connective tissue on the outside. This pretty packaging separates it from other organs, like a kind of garden fence. Under this, there's a layer of muscle known as the detrusor (yes, it sounds like an evil Pokémon to me, too). This muscle is wrinkled, and expands according to how full it is so the bladder can get bigger without tearing.

When it's empty, the bladder is more of an oval shape, and sits almost like a little bowl in the pelvis. As it gets fuller, it also gets rounder. When it eventually fills to the brim, its shape is more like a pear, or a balloon filled with water held from the top. The capacity of the bladder varies from person to person, but as a general rule, we ladies can store 350 to 550 ml, while men can hold a full 550 to 750 ml (yes, this is totally unfair).

On the inside, the urinary bladder is lined with a protective mucous membrane called the urothelium. Little stretch sensors sit here that measure how full the bladder is and pass this information on to the brain. I find these things really cute, and picture them as tiny smiley faces having fun in my bladder.

The internal mucous membrane, aka the mucosa, also

seals the bladder, stopping urine from seeping into the abdomen around it — like the tiles in a swimming pool. At the same time, the mucosa protects the bladder wall from trespassers, like bacteria and viruses, and ensures that urine doesn't come in direct contact with the bladder wall, which would be really painful and could cause infection and inflammation (more on this on page 61).

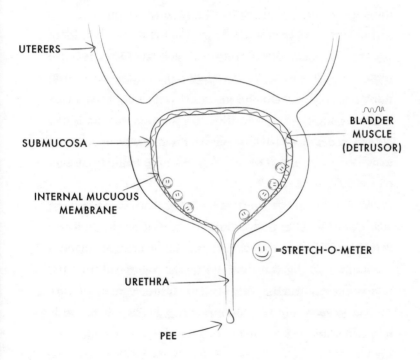

UTERERS

BLADDER
MUSCLE
(DETRUSOR)

SUBMUCOSA

INTERNAL MUCUOUS
MEMBRANE

=STRETCH-O-METER

URETHRA

PEE

The Urethra:
Wee-Slide to the World Outside

The urethra drains away the urine. It is attached to the bottom of the bladder like a little waterslide, and travels through the pelvic floor before finally emerging — at least in us women — in the vulva, just behind the clitoris. The female urethra measures roughly three to five centimetres, while fellas are endowed with an impressive 20 centimetres.

On the outside, the urethra is shrouded in a layer of muscle, which is in turn connected to the bladder and pelvic floor musculature. On the inside, it — like all organs that transport urine — is lined with a mucous protective layer that is supposed to keep bacteria out. If this doesn't work, our old nemesis cystitis will soon step into the ring (more on this in Chapter 3, page 61).

While the female urethra acts solely as a waterslide for wee, there's another bodily fluid that also exits the body via the male urethra: sperm. In men, the seminal ducts flow into the urethra, so this is also where sperm is excreted from the body — or ejaculated, rather. And for anyone who has always wanted to know but was too afraid to ask: no, it is not anatomically possible for men to urinate and ejaculate at the same time.

The Urethral Sphincters, The Body's Bouncers: 'You're not getting in … I mean out!'

To make sure everything is nice and watertight and that nothing drips out without you wanting it to, the urethra also has two sphincters, one mounted at either end. They act like bouncers, except that they're supposed to stop things from getting out, rather than in.

The internal urethral sphincter contains a network of veins in erectile tissue, and lies directly between the bladder and the urethra. When the internal sphincter is activated, the urethral mucous membrane contracts so that the bladder can fill up without dripping — kind of like a clamp that seals up your bladder until it is full and ready to be emptied. The external urethral sphincter is right at the bottom of the urethra, bordering on the pelvic floor. It is the final frontier to the outside world, so to speak, and, unlike the internal sphincter, we can consciously contract and relax the external one.

When the bladder is full, the internal sphincter opens so that some urine can flow down into the urethra. This reduces the need to pee a little, giving us more time. When we are ready to pee, we actively open the external sphincter and it's time to say, 'So long, urine, safe travels!'

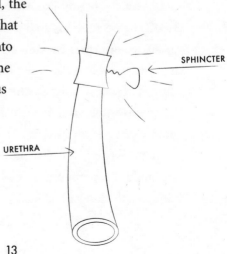

SPHINCTER

URETHRA

The Pelvic Floor:
The Mighty Sling Holding Everything Together

The pelvic floor is a complex muscle with three muscular layers woven together like a grid. It stretches in a gentle curve from the pubic region all the way back to the lower part of the spine. If it were a person, it would be the supportive kind who says 'Yes' to everything, and happily takes on any task assigned to them. A thank you? Don't mention it, the pelvic floor was happy to help. What exactly does it actually do, though? Well, first of all it sits at the bottom of the pelvis like a sling holding our internal organs in place. As if that wasn't already good enough, it also enables a straight, confident posture, which of course makes us look even more attractive (hey, sexy!).

Speaking of attraction: the pelvic floor is also important for a fulfilling sex life. When the pelvic floor muscles are fit, they can tense and relax properly, so you can have more fun in bed, including more intense orgasms.

The pelvic floor also ensures the sphincters work as they should by closing off the urethra so the bladder can collect enough urine. Only when the receptors give the go-ahead for evacuation does the pelvic floor let loose and relax the sphincter, allowing urine to flow away. Once the evacuation has been completed successfully, leaving you happy and relieved (in the truest sense of the word), the pelvic floor tenses again and the whole game can start over.

A fit pelvic floor also protects you from incontinence. Obviously, the more control you have over the external urethral sphincter, the longer and more securely you can hold in urine.

If, on the other hand, the pelvic floor is too weak and poorly trained, it often makes it hard to hold on, so even the smallest cough or sneeze can lead to a little stream of urine escaping.

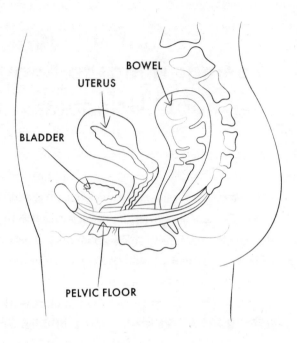

2.

Tinkling, Piddling, Peeing: Things to Know about Urinating

Some people read their favourite magazine on the toilet, while others do a Sudoku, go on Tinder or pick their nose (perhaps simultaneously). And me? I think about what is actually happening in my body while I wee. It's far more interesting.

Opening the Floodgates: How Peeing Works

You finished your litre bottle of soft drink about two hours ago, and now you're wondering precisely what has happened to the liquid in your belly. The drink found its way to your stomach via your oesophagus, and from there to the gut. Part of the drink has already been absorbed into your bloodstream through the intestinal mucosa and is being pumped around your body, where it supplies nutrients and absorbs toxins. That blood reaches the kidneys, which filter and purify it, and eventually turn it into urine. Now that urine is making its way to the bladder via the ureters.

The cute little stretch sensors we met earlier, on the urothelium, measure increasing tension in the bladder wall. Once it reaches a certain point, they send a message to the brain: 'Hey, Boss! We're starting to get full, please head for the next toilet.' At first this signal is so weak that many people don't really notice it. You usually only become actively aware of the signal when the bladder is already about half full, and a stronger urge tends to develop when the bladder is about 70 per cent full. You could go to the toilet at that point, but you don't absolutely have to. It's only when it reaches 90 per cent capacity that you get that typical I-really-have-to-go-to-the-toilet-RIGHT-NOW feeling, and should hurry to find a ladies' room.

A nice little fact that you might find comforting: the kidneys stop producing new urine when the bladder is 90 per cent full.

The bladder's entire process of storage and dispatch is controlled by the autonomic nervous system. The word 'autonomous' is a giveaway here: you can't influence it consciously because it's basically on autopilot. The autonomic nervous system is divided into several branches, two of which are important for our bladder: the sympathetic one and the parasympathetic one. (In case you were wondering: the autonomic nervous system itself is one of two branches of the peripheral nervous system (PNS) — *i.e.* all the nerves and ganglia outside the brain and spinal cord; the other branch of the PNS is called the somatic nervous system.)

The sympathetic nervous system is the active part of the autonomic nervous system, and is responsible for the bladder being able to collect urine properly. It does this by keeping the internal urethral sphincter muscle closed while simultaneously relaxing the bladder muscle. Once the maximum capacity has been reached, the brain gives the go-ahead to pass water, and the micturition reflex is triggered. This stage of the process is managed by the parasympathetic nervous system, which is sometimes called the 'rest-and-digest' system. It does exactly the opposite of the sympathetic nervous system: it contracts the bladder muscle while simultaneously opening the internal sphincter, which further increases the urge to pee. And all of this happens without you consciously controlling a thing!

Don't worry, though: you won't just go ahead and wet your pants then and there — you are of course equipped with a second urethral sphincter. If you tense the external sphincter, which you can control yourself, the need to pee can be staved off for a while longer. Once you've found a suitable evacuation site, you actively relax the external sphincter and the bladder muscle contracts even more, squeezing the urine out. And voilà!

Your soft drink has taken about two hours to go through this entire production process. And even though you feel relieved afterwards, the bladder is never completely empty; on average, up to ten millilitres of urine always remain.

A fun little fact: you hold your breath when you pee. Why? To exert pressure on the bladder and make it empty more quickly, tension is required in the abdominal area. The diaphragm helps apply this pressure, which interrupts our breathing for a moment. Once everything is going swimmingly, you can start breathing normally again. This process is so routine that you don't even notice it.

Sitting Pretty: How to Go to the Toilet Properly

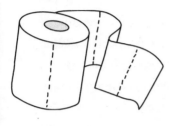

Giulia Enders' book *Gut* has already taught us the best way to go about our number twos. There are also a few things to keep in mind when it comes to number ones, though. Yes, I know. You just want to get it over and done with quickly, so you never really think about it. As with so many things in life, though, mistakes can be made.

First off, a tip I got from a urologist. It might seem obvious, but I never knew: wash your hands before you pee. Just think about the things you touch over the course of a day! And then you go and handle your genital area with those grubby little paws? It's no wonder bacteria, viruses or fungi sometimes get lodged there.

Now that your hands are clean, let's get started. Sit as comfortably as you can on the toilet. Your feet should rest gently on the ground, parallel to each other. Don't press your knees together; instead, hold them about hip-width apart in a natural position. Your upper body should be straight, but make sure that you don't lapse into a swayback posture. Better yet, lean forward a little. Just a tad. This way, you can ensure your urethra is in the optimal position to pass water: the urine won't have to journey through any unnecessary curves or kinks, and everything can flow smoothly.

Most of your body weight should be resting on the rear part of your pelvic floor. Meanwhile, you can do whatever you want with your arms. Stretch them out to the sides, put them on your head, do Madonna's 'Vogue' dance: anything's allowed, so long as it relaxes you. I recommend just resting them normally on your thighs. Once you're relaxed and sitting in the correct position, you can move on to urination itself.

Please don't push or squeeze like a world champion. This is not actually a pissing contest, and nor is it a speed test. Relax and let things take their course. If you do push, you won't actually speed up the peeing process; instead, you will be pushing the bladder, uterus and rectum down, which can damage the pelvic floor in the long run and cause organs to prolapse. So don't rush it.

Once the golden fountain starts to run dry, it's important to finish in the right way. The same thing applies here, though: please don't push, even if it feels as if the last bit just has to be squeezed out. It's better to wait a little and let it drip until nothing else comes out. Then reach for some loo paper and — this is super important — wipe from front to back. Otherwise, there's a danger that gut bacteria from the anus will find its way to the urethra, where it can unleash a nasty UTI.

To ensure you're well equipped for your next trip to the toilet, here's a summary of proper pee etiquette.

PROPER PEEING:

- Wash your hands!
- Sit in a relaxed position
- Place your feet flat on the floor, knees parallel and hip-width apart
- Sit up straight, bending slightly forward
- Your weight should be at the back of the pelvic floor
- Never (never!!!) push while peeing
- Wait until the last vestiges have dripped out
- Wipe from front to back with toilet paper
- Flush!
- And again: wash your hands.

Why Do We Get Thirsty and Need to Drink, Anyway?

Before we go any further, let's enjoy a nice big glass of water. Cheers! Drinking enough is enormously important. That's because humans are made of more than 50 per cent water. We don't just carry this water around with us, saving it for hard times — we're not camels. Our bodies need it for all sorts of tasks: it powers our organs by transporting oxygen and nutrients through our blood into our cells; it helps get rid of toxins by carrying metabolic products to the organs of excretion; and, thanks to the cooling power of sweat, it ensures that we maintain a constant body temperature. It's super important that our water tanks don't run dry.

To keep everything functioning at peak condition, you should drink at least 1.5 litres of water each day. And not just when you notice that you're thirsty; this feeling only crops up when you are already drying out and your body is running on fumes. Thirst is triggered by special sensors located in the diencephalon (part of the brain also known as the interbrain). When the body is thirsty, an antidiuretic hormone — ADH

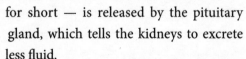

for short — is released by the pituitary gland, which tells the kidneys to excrete less fluid.

And what happens to the body when you don't have enough to drink? Firstly, the blood becomes stickier, so it flows more slowly. Like porridge in a pot, it gradually gets thicker and less fluid. Nutrients and oxygen take longer to

1.5L

reach the body's cells, and waste products are excreted more slowly. You get sluggish, lose concentration, and grow tired and irritable. If you still refuse to reach for the water bottle, your body will thank you with an extremely dry mouth, which — as you may know — leads to pretty gross breath.

If you still don't drink anything, you won't just annoy your fellow humans with your bad breath, it'll also lead to headaches, problems with your circulation and increased body temperature. In extreme cases, it can even result in delirium and fainting.

Once you've realised that you really must bring your fluid levels back into balance, it's not enough to quickly down a bottle of water in the hope your body will recover in ten minutes. It can take a day until everything is functioning normally again, your headache has disappeared and you're back at the top of your game. So, to prevent all this from happening in the first place, you should always have a bottle of water by your side.

Of course, it goes without saying that you should drink more in the heat or when doing physical exercise, but fever, vomiting, and diarrhoea along with foods rich in salt or protein also increase fluid requirements. And if you're thinking, 'Whatever, I just sit at my desk in the office all day and don't really move much …' You still need to have plenty to drink. Even when at rest, we lose around 2.5 litres of water each day. Half a litre of this is just through breathing. We pee out 1.5 to 2 litres, and the rest is lost through faeces and sweat.

Going to the Toilet:
How Often is Actually Normal?

It's a very important question — but unfortunately it can't be answered definitively. Every person is built differently and no two bladders are alike, either. In general, though, it's true to say that the bladder voids around 1,000 to 2,000 ml of urine each day, and each time you go to the toilet, you pee away about 200 to 400 ml. How often you need to go to the toilet depends of course on how much you drink, but if you dutifully drink your 1.5 to 2 litres each day, doctors suggest you should be going to the toilet at least four times a day.

If your bladder forces you to go to the toilet a lot more than this and your body is producing more than 2,000 ml of urine a day, experts call it polyuria. If only 400 to 500 ml make their way into the bowl a day, you're dealing with the medical condition oliguria (aka low output of urine). If you pee even less than that — less than 100 ml a day — you're suffering from anuria.

Great, we've now laid out the groundwork. But there are of course other questions that are important, such as:

- What have you had to drink?
- What have you eaten?
- Have you been physically active and perspired?
- Are you physically healthy in general?

The fact that you need to go to the toilet more often when you drink more is pretty obvious. What exactly you've had to drink is also important, though, because some drinks stimulate the bladder. They are known as diuretics; that is, they promote the production of urine. Coffee does it for me. As do green tea and cola. And why do these drinks lead to more frequent trips to the toilet? Because it's not just your circulation and brain that they wake up, spurring them on to perform overtime; they also awaken your dear old kidneys. The caffeine means that the kidneys are supplied with more blood than usual, so they work more quickly and produce urine faster.

As well as drinks, there are also solid foods that have a diuretic effect, *e.g.* asparagus. In the case of asparagus, this is partly because it is 90 per cent water, and so it fills up your liquid reservoir quite nicely. It also contains asparagusic acid and potassium, both of which stimulate renal activity and thus the production of urine.

Avocados, tomatoes, cucumbers and oranges contain potassium as well, so they can also be pee stimulators. Once again though, it depends how your body processes the food. There are apparently even people who say they have to go to the toilet more often when they eat rice.

How much you've sweated over the course of the day is also important. If you lose a lot of water this way, the kidneys need to reduce the production of urine so that the body retains enough water to keep functioning properly. So, the more you sweat, the slower the bladder will fill up, and the less you need to pee. The loss of fluids through sweat is why it's particularly important to have plenty to drink when doing exercise or when it's hot.

Another important factor is your health. Are you perhaps on medication or dealing with an underlying illness? There are a few conditions that make you dash to the toilet more often, such as diabetes, an overactive thyroid or heart failure. Your doctor can check for all this through urine and blood tests and treat them if necessary. On the other hand, you might be taking medication that has a dehydrating effect and increases the urge to urinate. Check the side effects on the package and consult a doctor or pharmacist if you have any questions.

Too Much Tinkling?
Why You Might Have to Go More Often

Do your colleagues affectionately call you the Whizz Kid? Do you know the bathroom attendant at your favourite restaurant by their first name, and always know where the next (usable) public toilet is? Then you probably have to go to the toilet a lot. But why?

Some people just have to go more often than others. Doctors call it polyuria (see above). If there is no medical cause, though, having to pass water frequently doesn't have to be a bad thing. In fact, it's entirely the opposite: if you pee more often, the urinary tract is flushed out more often, and bacterial infections crop up less often. If you are one of those people that just drinks a lot, it stands to reason that you'll have to pee more often than others. On the other hand, you might have to urinate frequently for far less pleasant reasons.

You're anxious

You've got a job interview. You're giving a speech, or you have an important meeting. Of course it would be now that your bladder decides to act up every ten minutes. Don't be too hard on it, though: it's under enormous pressure at the moment, and it might not be as good at playing it cool as you are. Because the bladder is controlled in part by the central nervous system, which includes the spinal cord and brain, she is also influenced by fear, tension and stress. There's more on why job interviews, etc. send us to the ladies' room so often in the section 'The Bladder as the Mirror of the Soul' (on page 189).

You have or had a UTI

UTIs usually announce themselves by making you need to go to the toilet more often and making it really painful. Sometimes, though, the pain is absent, and the 'only' symptom is the increased frequency. Recognising this complaint as a bladder infection is not at all easy.

So, if you notice a change in your habits of micturition, *i.e.* urination, even though you're drinking the same amount as before, you should monitor it and have it checked out by a doctor if necessary. And be aware: even after the infection has subsided, your bladder may still send you to the toilet more often. If this happens, it's to do with what's known as a

27

hypochondriac fixation — in other words, it can be a purely psychosomatic complaint. While you were sick, you thought about going to the toilet so much and were so afraid of being permanently terrorised by your bladder that now you really do need to go to the bathroom more often.

You feel unwell and are under too much stress

There's an expression in German that can be translated as 'the bladder cries out'. Just like the skin, gut and hair, the bladder is also a mirror of the soul and can make you aware that something's up. Is there too much stress in your life? Are you trying to cover up the fact that you're no longer coping?

Your bladder can reveal all this to you through hyperactivity. The pressure you're feeling can make the tissue surrounding the bladder or the entire pelvic floor cramp up, causing the bladder to contract more often. The right relaxation exercises can help here. If they don't, you should think about paying a visit to a psychologist. Perhaps your bladder is acting this way to try to tell you that something in your life needs to change.

Your bladder is a workaholic

This is when you suffer from an overactive bladder. Unfortunately, there hasn't yet been much research into why lots of bladders suddenly start pulling all-nighters and going into overdrive. Most of the time, though, it's because the receptors on the bladder wall that notify the brain that it's full aren't working properly. They sound the alarm at a really low volume of urine, so even when your bladder is only half full you feel an uncomfortable, unwarranted amount of pressure. And yes, under certain conditions this pressure can be so strong that you (or more precisely, your sphincter) can't stand it any longer, and some urine escapes.

If you have the feeling your bladder might be overactive, you should definitely have it checked out by a specialist. She'll be able to perform various tests to assess whether and why exactly your bladder has turned into a workaholic, and what you can do about it (more on this on page 86).

You're pregnant

Uh-oh, when was your last period? And have you been taking proper precautions? Now, I don't want to make you anxious about an unplanned pregnancy, but frequent urination really can be the first sign of being in the family way.

Why? Well, firstly this is down to the hormones that make your body ready for a baby. A lot of the sex hormone progesterone is released at the beginning of pregnancy to support the implantation of the zygote, *i.e.*, the fertilised ovum. This hormone affects the machinery of the bladder by relaxing the muscles. The organs are all supplied with more blood, too, so the kidneys produce more urine. You can read more about this on page 169.

Your hormones have gone haywire

Can you think of a single bodily phenomenon that you can't blame on your hormones? Hmm, tricky! Make a point of noticing whether you need to go to the toilet more often before, during or after your period. As in pregnancy, progesterone is released at this time, which can affect the bladder.

If you're already going through the inconvenience of menopause, frequent urination could be due to your body producing less oestrogen. This makes the bladder more susceptible to irritants in the urine, which in turn makes you need to go to the toilet more often. In this case, you can find respite through hormone suppositories that release their active substances in exactly the right places, *i.e.* the urethra and the bladder.

You have diabetes

Unfortunately, a more serious illness can be lurking behind the 'harmless' need to go to the toilet every ten minutes: diabetes. If you also feel tired, run-down and have lost weight, you should get it checked out by your doctor, since one of the classic symptoms of diabetes is increased urination. It won't take long to get answers through a blood or urine test.

You're taking certain medications

Some medications deliberately increase urination, such as those prescribed for high blood pressure or oedemata. Aside from this, you might be taking medicine that contains diuretic ingredients. It's best to have another careful read through the packaging.

Foods That Excite the Bladder and Fuel the Urge to Pee

You are what you eat … and you tinkle what you eat, too! There really are some foods and drinks that fire up your urge to urinate. This can be pretty helpful if you're suffering from cystitis or another UTI and want to flush the bacteria out of your body quickly. On the other hand, it is of course pretty annoying having to vanish to the loo every twenty minutes.

So that you know your enemies (or friends), I've listed some of the most common culprits below.

Coffee, cola & co.

Coffee might help you stay awake, but it's your enemy on long drives when you can't take a toilet break. I don't think there's anyone who can easily gulp down three cups of coffee without having to go to the toilet every half hour after that. But why, exactly? While it's true that caffeinated drinks are often diuretic, it's not true that coffee removes more water from the body than it gives it. People used to think this was the case, but it's a myth. Caffeine does, however, stimulate the stress hormones adrenaline and cortisol, which make the heart pump faster, the pulse increase and the blood vessels widen. All this leads to the renal vessels (the ones that go to and from your kidneys) widening and working more quickly and effectively. The result: more urine is produced, and more quickly.

31

Carbonation

Ok, ok, carbonation isn't exactly a drink, but I can't leave it off this list of diuretics. Why? Fizzy drinks, or 'tingles', as my sisters and I always used to call them as kids, irritate the bladder's mucous membrane. If you've already got an irritable bladder that sends you to the can more often than you'd like, you're better off ordering still water rather than sparkling.

Alcohol

You've barely downed your second glass of prosecco when, alongside feeling a bit of a buzz, your bladder starts reporting for duty. What? Already? There can't possibly be that much in there yet. But, remember that alcohol is technically a toxin, and our bodies and above all our kidneys are real purists and want to get poisons out of their system as quickly as possible. So the urge soon grows. Alcohol also often has bubbles in it, which, as explained above, increases the need to go.

The real culprit, though, is ADH (the antidiuretic hormone), or rather its absence. ADH is produced in the brain and regulates the water balance in the body. As described in Chapter 1, the kidneys draw water out of primary urine and give it back to the body as part of the filtration process. When you drink alcohol, though, the production of ADH is put on ice. The consequences? The kidneys reclaim less water, so more ends up in the secondary urine, *i.e.* the urine that's ready to be peed out, and the bladder fills up. This urine is really diluted, which is why alcohol pee looks kind of white and transparent instead of golden yellow. It takes a while for your body to notice that there's not enough ADH around, and this is why the urge to pee strikes all the more suddenly and fiercely when you're out at a bar (where, naturally, there is an enormous queue to use the ladies').

Are you familiar with the terrible piddle loop? The merry-go-round that you hop on as soon as you give in to the first urge to pee while under the influence? Is it true that you should resist the first call of nature as long as possible so that you can avoid having to run to the WC every twenty minutes after that? Surprisingly, it is! The cycle works like this: through the increased consumption of alcohol, the brain, along with the autonomic nervous system, which controls the bladder, enters a boozy slumber. The brain does in fact notice that the bladder is getting full, but doesn't get as far as sending out a direct announcement telling you to empty it. If you do so ASAP anyway, you remind your body of the bladder's existence and the brain orders it to get back to work. It's like it's saying, 'Ah yes, that's where we were …' You set it off, so to speak, and make the bladder need to empty itself regularly again. Do yourself a favour and delay going to the toilet for as long as possible when you're drinking alcohol.

Spicy food

Do you find normal levels of spice too boring and always order dishes extra hot at your local Indian or Thai restaurant? Then be happy that your bladder isn't sitting at the table with you, because it hates spicy food. Generally speaking, really spicy food acidifies urine, which irritates the bladder's mucous membrane and makes you need to go to the ladies' more.

Vegetables like asparagus, cucumbers, and avocado

As already mentioned, these kinds of vegetables contain quite a lot of water, so we're ingesting fluids when we eat, and they also provide us with a lot of potassium. Potassium is important for the distribution of water in the body, so consuming it can have a diuretic effect.

Sugar

So, sweeties, you're going to have to be strong. Sugar is delicious, but it isn't exactly good for the urinary balance sheet. If your blood sugar levels shoot up, your body will try to excrete the excess sugar through your urine, so it's guaranteed that after a sugar fest you'll need to go to the toilet more often. This in turn makes you thirstier, so you fill up your water tank … and welcome to the vicious toilet circle! The good news is that you really have to scoff a LOT of sweets, chocolate or biscuits before this happens.

Foods that make your urine more acidic

When your urine is feeling really sour, it lets you know by sending you to the toilet more often. The more acidic your urine is, the more it irritates the bladder wall, making you need to go to the toilet more often and more urgently. Your urine is made acidic by protein-rich foods such as cheese, eggs and meat, along with pasta, bread and cola. If you need to go to the toilet frequently, perhaps you should take a look at your diet and consider adjusting it — you can ask a nutritionist or dietician for more guidance.

Drinks That Fill the Bladder Up More Slowly

Looking for drinks and snacks for the cinema or a long drive? Researchers at the universities of Loughborough, Stirling and Bangor in the United Kingdom have addressed this question. They had 72 human guinea pigs consume a litre of a liquid — such as milk, cola, orange juice, ice tea, hot tea, or an electrolyte drink — over the course of 30 minutes to test whether certain drinks were excreted more slowly. The winners: electrolyte drinks and milk. These drinks stayed in the body for nearly 1.5 hours longer than did cola and co. Orange juice also performed well, only triggering the urge to pee later. Researchers hypothesise that the reason for the fluid being kept in the body longer is the specific composition of the drink. Orange juice, for example, contains vitamins and fruit pulp, which take longer for the gut to break down and process, thus taking longer to arrive in the bladder.

Vitamin B

As we've seen, the urinary tract works closely with the autonomic nervous system, which is where decisions are made about when the bladder should be emptied. To optimise the performance of this nervous system, your body needs Vitamin B more than anything else. To stock up on it, try including more foods such as wholemeal bread, berries, walnuts and sunflower seeds in your meal plan.

Why Women Need to Go to the Toilet More Often Than Men

I'm not a big fan of clichés, especially when it comes to the differences between men and women. I'm sorry to say, though, that it really is true that women need to go to the toilet more often than men. This is entirely down to our anatomy. Our bladders are a lot smaller than men's — about a can of cola less, give or take — because they have to make room for the uterus.

Cystitis can be another reason for women's more frequent tinkling, though, since one of its annoying symptoms is an increased urge to urinate. And, because the female urethra is also shorter than the male (remember: men's measure 20 to 25 centimetres, while women's are only three to five), it's much easier for bacteria to sneak into the female bladder and make trouble.

How Small Children Learn to Control Their Urge to Pee

This heading could just as easily be 'How Adults Learned to Get to the Toilet in Time'. If we hadn't, we would all regularly pee our pants or need to walk around with nappies or catheters.

In the first few months of life, babies pee roughly thirty times a day. From the age of three, *i.e.* by the time they can (almost) go to the toilet all by themselves, they 'only' go ten times a day. Interestingly, babies become aware that they need to pee quite early on. Until the end of the 1990s, people still thought that babies' bladders were too small or immature to work properly, and that they over-flowed when they were full; or that babies didn't even register when the bladder was getting full and every-thing passed through it. But thanks to more recent research, we now know that in babies, just as in adults, the autonomic nervous system regulates the urge to urinate.

You can sometimes observe this fact in real time: some babies make strange noises before they have to pee, while others move in funny ways (my nephew, for example, twitches his little legs very sweetly before he has to go). Apart from this, it has been found that babies actually stay dry during slow-wave sleep — so their bladders must be sleeping, too. It's only in the lighter sleep phase, just before falling asleep or waking up, that the bladder becomes active and the nappy needs to show what it's made of.

Children can indicate, or even say, that their bladder is full from when they're about one and a half years old. This is when the feeling of urinary urgency develops. About a year after this, *i.e.* at the age of about two and a half years, they can normally manage to hold their pee in until they've reached a toilet. From this point on, kids can actively control their external urethral sphincter. Children can usually go to the toilet entirely on their own, including doing everything that goes along with it — lifting the toilet seat lid, pulling down their pants, etc. — when they're about three. From then on, they can resist the urge to pee for longer or even go to the toilet as a precaution, for example before a long drive.

While it's true that the connection between the bladder and the brain is in operation even in babies and small children, it does have to be switched on properly and allowed to mature. It takes time to learn to deal with all the hubbub of the body, to interpret it correctly and then to react appropriately. Hearing the bladder's signals and understanding them properly isn't that easy. After all, they're very timid and quiet at first, so you can ignore and then forget about them. Especially when you're a baby and everything around you is super new and exciting. Many children only become aware of and announce their need to go to the toilet when their bladder is full to the brim and hammering on the panic button. Unfortunately, by then it's often too late and their trousers are wet!

When exactly each child gains complete control over her bladder depends entirely on the individual. The brain is the boss here, so it can't really be taught or forced, either. What parents can do, though, is show — with love and patience — how going to the toilet works. In other words, how you find the toilet, how you behave there, roughly how long the entire process takes, and so on. And, really importantly: punishments and threats will achieve the complete opposite of their desired effect.

It can take children a while to get the hang of all this. If a child hasn't managed to gain full control over their urinary equipment and stay dry after the age of five, it should be looked into by a doctor.

While you were sleeping: bed-wetting aka enuresis

When children over five years old wet themselves in their sleep, it's called enuresis. There is no overarching reason why some children still wet the bed after a certain age, but enuresis is often understood as a dysfunctional stress management technique used by a body under psychological strain. For example, children might continue to wet the bed past the age of five if there is a lot of conflict at home, or if their parents don't pay them enough attention. Enuresis can also have physical causes, though. For example, the capacity of the bladder might be too small, so it just can't hold very much fluid. It's also not uncommon for the nerves necessary for micturition not to have developed properly yet, so they might be sending false signals. If there are issues with nerve development, the older the affected child or young person is, the more frequently they will wet themselves at night. And, of course, the older they get, the more uncomfortable and embarrassing it is for them! This is why it's important to find a paediatrician early on who can recommend various treatment options.

Enuresis can also occur in adults. There's more on this on page 164 under the topic of incontinence.

OMG: Can Your Bladder Really Burst?

'Seriously, I need to go to the toilet so badly my bladder's about to burst ...' We've probably all moaned like this at one time or another. In the movie that plays in your head, your bladder then explodes with a big bang, filling the entire abdomen with urine. All that remains is a tiny little shrivelled bladder, like a burst balloon. And ... cut!

But is it possible? Can the bladder really burst if it's not emptied? Apparently, it did happen to the Danish astronomer Tycho Brahe. In the 17th century, Tycho was invited to a banquet held by Emperor Rudolf II. The problem was that pompous events such as this always followed strict court etiquette, and you couldn't just get up and go to the toilet whenever your bladder called. This is exactly what Tycho's did, urgently. The end of the story: Tycho died a short while later in indescribable pain — purportedly of a burst bladder. Various scientists later examined the case and found out that Tycho Brahe was more likely the victim of uraemia than of his overfull bladder. (For the nerds amongst us: uraemia is when there are high levels of toxic substances in the blood that the kidneys usually eliminate via urine. This happens when the kidneys aren't able to purify the blood well enough, usually because of severe kidney disease. These harmful substances eventually poison the body.)

I can reassure you (and myself) that the bladder can't actually burst. Before that happens, it empties itself of its own accord. From a volume of between 400 and 1,000 ml, your bladder can't take the pressure anymore and inevitably gives in. The collected urine simply leaks out and we wet our pants. It's unpleasant, sure; but it's still better than a burst bladder.

Ruptured Bladders

The bladder cannot explode, but it can tear. Before you despair: this only happens very rarely, and not because you've simply held on for too long. What's known as cystorrhexis, or a rupture of the bladder, usually occurs because of external impact, that is, when extreme pressure is suddenly put on your bladder, *e.g.* in a car crash. The probability of the bladder rupturing or tearing is, by the way, much higher when it is full than empty. The fuller it is, the more fragile.

When your bladder tears, the Hollywood scene described above really does play out inside your body, and urine flows into the abdominal cavity. It doesn't happen in such an action-packed way, though; it's much slower and more inconspicuous. The most common symptoms of a ruptured bladder are pain in the lower abdomen, bloody urine, frequent passing of water, or even that the tap runs dry. And how do you patch up a ruptured bladder? The bladder can usually heal small tears itself. While this is happening, a catheter is often used to relieve the pressure. Larger tears must be surgically sewn back up.

Paruresis, aka Bashful Bladder: The Fear of Public Urination

Imagine the following scenario: you're out and your bladder has just given your brain the absolute last warning that it needs to be emptied. A toilet had better show up right now, or else. Then, when you've finally found a loo and are ready to go, only a few drops or nothing at all comes out. What's going on here? Is your bladder still asleep? Did you perhaps not need to go so badly after all? No, your bladder is just shy. It might sound cute, but I mean it in all seriousness.

The phenomenon of the shy bladder really does exist. The technical name for what I affectionately refer to as pee shame is 'paruresis'. It's when a full bladder does not want to be emptied when others might be nearby, so it goes on strike. This often happens in public toilets. This is not due to disgust or because of fears about bacteria, but rather out of anxiety that other people who are visiting the toilet at the same time might hear it. People who are ashamed of the noise of their pee, that is, the splashing of urine on water, are not at all uncommon (you'll notice them: they might turn the tap on for a strangely long time, or suddenly start blasting out music). Why some people suffer from this stage fright isn't really known. Paruresis has only been studied as an actual condition since 1980.

The International Paruresis Association estimates that 7 per cent of the population has a shy bladder.[1] Paruresis can exist in various degrees of severity. Some people just need more time and concentration until they get into the flow of things and can pass water. Those with a more severe phobia cannot tinkle in public places at all: their bladders go completely on strike even though they may be full to bursting, and not a single drop leaves their urethra. It's really uncomfortable.

As to why some bladders are more shy than others, there's no one-size-fits-all answer, and, if you have problems peeing, then physical causes such as bladder stones and tumours must first be ruled out. Most often, though, the problem is psychological.

To better understand the fear of public urination, we need to get to know the autonomic nervous system better. Remember, we cannot consciously control the autonomic nervous system; its processes happen automatically without us even realising. Not only is it responsible for the urinary tract, it's also in charge of our digestion, the production of sweat, and the dilation of our blood vessels, for example. It's what makes us go red when we're embarrassed or get sweaty hands when we're nervous. Can we control or suppress it? No, unfortunately not.

Let's return to those two branches of the autonomic nervous system, the sympathetic one and the parasympathetic one. The sympathetic nervous system is the active part that ensures we get properly into gear. It makes our hearts beat faster, dilates our pupils and slows down our digestion to avoid wasting energy. The parasympathetic nervous system, on the other hand, is a real anchor of calm. It's responsible for relaxation and comes alive when we're chilling on the sofa after a hard day. When it is activated, your heartbeat and breathing calm down and your metabolism becomes slower.

And what does all this have to do with a shy bladder? Well, in order for the entire peeing process to work, the autonomic nervous system has to operate smoothly, and the sympathetic and parasympathetic nervous systems must be perfectly in tune. The former ensures that the internal sphincter stays tightly closed while the bladder is filling up, so nothing drips out. The parasympathetic, meanwhile, is responsible for the bladder muscle contracting and the internal sphincter opening so urine can be let loose.

In order to pee, the sympathetic nervous system needs to dial down. But it is triggered by stress and commotion. So, if we perceive going to a public toilet as an extremely stressful situation, the sympathetic nervous system holds the sphincters closed, even though it shouldn't do so at all.

There can be many reasons that people have a phobia of a particular toilet situation: bad experiences in childhood, peer pressure, fear of comparison — the list could go on. Since paruresis is an anxiety disorder, though, it can't be fixed simply through medication or surgery. You can, however, get on top of it quite well through methods such as psychotherapy — *e.g.* cognitive behavioural therapy, which, rather than working on the fear of the toilet or of urination itself, will help you look at why you are ashamed of being observed or judged — or relaxation techniques.

What Happens When You Hold Urine In for Too Long?

It's half past ten in the morning and your bladder is piling on the pressure. It feels like it's been doing this for two hours already. But you can't vanish to the toilet right now — you don't have time. Now we know that the bladder won't explode, but it's still not healthy to resist for too long. If you occasionally put off going to the toilet when you're at the cinema or in meetings, no problem; if, however, you keep doing it over a longer period of time, for months or even years, you're conditioning your bladder. It gets used to storing as much urine as possible by expanding more and more. This overextends the bladder muscle and wall, until at some point they can't contract properly anymore. It's like a scrunchy that you're constantly doubling and tripling, wrapping it around and around your ponytail (women with thin hair who can't get a proper ponytail — I feel you) until all the tension has gone. At some point, this scrunchy will grow loose and unusable.

If the fibres responsible for expansion become overextended too often, the stretch sensors that tell the brain how full the bladder is can also stop working properly. The result: it's hard to pass water. This is firstly because the brain is notified of the urge to urinate later, so the person in question needs to go to the toilet much less frequently. Secondly, it's because the pressure needed to pee properly fails to materialise. The bladder is too big, too flabby — yep, it's just got too lazy. This is why this disorder is also called lazy bladder. The danger of residual urine is also high here, because the bladder no longer has the power to evict all the pee. (For more on what makes residual urine dangerous, see pages 111–12.)

You might know the overextended, aka lazy, bladder by other names: teacher's or nurse's bladder. It got these names because the professions are so stressful and demanding that teachers and nurses just don't have time to go to the bathroom regularly enough. This means the bladder collects urine over the entire working day, grows bigger and bigger, and over time becomes distended. If this happens to you, urologists will know what to do. So don't shy away from discussing your problem.

Is it Bad to Go Too Often?

And what about the opposite: going to the toilet too often? That is, another quick tinkle before getting into the car. Or leaving the office. Or going to the toilet in a department store because you don't know how far the next one might be. What does just-one-more-quick-pee-or-I'll-regret-it-later, aka panic piddling, do to the bladder? And could it even be dangerous?

First of all, rest assured: your bladder won't shrink just because you're a more frequent pisser (for information on how your bladder can shrink, see pages 107–8). What can happen, though, is that you get used to going to the toilet more often. In other words, the problem is not an anatomical but a sensory one: you feel like you need to go to the toilet urgently, but your bladder actually still has plenty of room and could easily wait longer to be emptied. Despite this, you really feel like your bladder is full to bursting. And yes, if you don't go to the toilet, you will wet yourself.

You actually only have yourself to blame for this. By going to the toilet too frequently, by constantly going 'just in case', you've made your bladder accustomed to being emptied at any opportunity, even when there's not much in it. Your bladder is basically acting like a spoiled child that wreaks havoc if it doesn't get what it wants. You've taught it this bad behaviour, though, so you can teach it to behave itself again. To reprogramme the bladder, avoid going to the toilet too often. For tips on bladder training, see pages 94–5.

What Urine Is Made Of

Urine is the body's waste water. Its main components are water (95 per cent), electrolytes, waste products that build up through the body's metabolic processes, and impurities we have consumed through alcohol, medication and so on that should be removed ASAP.

As we've already learned, urine is produced in the kidneys, first as primary urine and then as secondary urine. Primary urine is still enriched with way too many useful materials, such as glucose, minerals, amino acids, proteins, and electrolytes, so as it passes through the renal tubules, the good stuff is transported back into the blood. What's left over is the 'secondary urine', which, along with all the water, is made up of lots of metabolic waste products — things like creatinine,

47

uric acid, urea and sugar. Experts often like to describe what kidneys do as 'filtration, reabsorption and secretion'. The stuff that needs to be secreted — the secondary urine — is channelled from the kidneys into the bladder by the ureters.

Beauty Pee

Urea for the skin

I'm sure I'm not the only one who's read it on some little jar of hand cream or another: urea. Say what now? Are we really smearing pee on our skin?

Urea, also called carbamide, is one of the main products accumulated when our livers synthesise proteins. It is mainly excreted through the kidneys and bladder, but also through the bowels and sweat. Your perspiration means you actually always have a thin layer of urea (or pee film) on your skin.

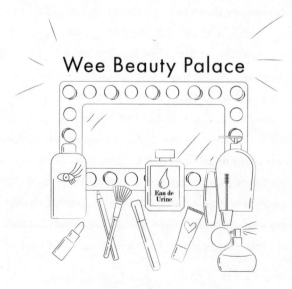

It might sound a bit gross at first, but it has a fantastic effect in terms of beauty: urea is a top-notch moisturiser. It attracts moisture in the air and stores it on the skin.

If you don't have any urea on your skin, it can dry out and get itchy or flaky. Creams with urea are really great for neurodermatitis or eczema because they provide the skin with natural moisture.

For anyone who feels deeply disgusted and is now throwing their creams, lotions, shampoos and so on in the bin — don't worry, you're not rubbing 'real' urea into your skin. Since 1882, urea for use in toiletries has been produced artificially from carbon dioxide and ammonia.

What's auto-urine therapy all about?

Urine therapy is a method of alternative medicine celebrated by its fans as a solution to skin ailments such as acne, neurodermatitis, psoriasis, allergies, osteoarthritis, rheumatism, and even asthma. Some leading representatives of natural medicine oppose it, though, and so far there is no evidence to support its health benefits.

I can't judge from personal experience whether or not therapies using your own urine are beneficial. But if you want to try it anyway, you really must ensure that the urine used is as free of germs and bacteria as possible. This means that this therapy is off-limits if you often suffer from UTIs, or if you regularly take medication. Apart from this, it's best to use the midstream of your morning micturition, as this is the cleanest. Midstream means: start peeing, then collect some in your container, and then finally let the last bit flow into the bowl again.

Beyond this, the therapy can take lots of different forms. If you suffer from neurodermatitis or have generally bad skin, you can smear urine directly onto the affected area and just let it soak in. To do this, pee into a sterile cup, dunk a cotton pad into it and then dab it onto the relevant part of your body. Don't worry: the smell will disappear. The hard-nosed really get their fill of urine by drinking it, which is supposed to strengthen the body's defences and make you fitter. But please, if you want to try this then only ever use really fresh urine. Collecting it in canisters and storing it in the fridge is not a good idea because it degrades. There are even people who have their urine administered intramuscularly, *i.e.* injected into their muscles. Crazy, right? Please don't try this. It hurts and can cause muscle inflammation.

Opponents of urine therapy warn again and again that urine is not free of bacteria or germs and can therefore make illnesses even worse — after all, the body just excreted toxins through this very urine.

Asparagus Urine:
You Pee What You Eat — Or Do You?

Why is urine smelly after you've eaten particular foods? Pee contains metabolic products that are influenced by the various foods we feed the body. This is why urine smells different from usual after you've enjoyed coffee, garlic, cabbage or onions. And, famously, it has a really strong smell after you've eaten asparagus. One spear of asparagus is all it takes for a slightly acidic, milky smell like sulphur to waft out of the toilet bowl. Asparagusic acid is the substance that causes the pong: the typical sulphur smell in urine develops when enzymes break it down.

Not all people find this, though, because not everyone has the enzyme that breaks asparagusic acid down into its stinky sulphuric components. The exact reason hasn't yet been established, but it's clear that this particular enzyme is inherited. So, if your urine stinks after eating asparagus, it's safe to assume that you have your parents, your grandma or your great-grandfather to thank.

If you're breastfeeding, you should also be aware that this asparagus aroma passes into mother's milk. So, if your baby flat-out refuses to take the breast, it could be because you've been eating asparagus.

What the Smell of Your Urine Can Tell You

Fresh urine smells neutral, that is to say, it doesn't smell of anything. It's only when it gets older and bacteria start breaking it down that the smell changes. It's exactly the same with sweat. Straight after an exercise class, your armpits smell really fresh; it's only after you've left the gym and arrived at your favourite café that they develop an aroma.

If your pee has — how shall I put this — a strong odour even when fresh, it could either be because you've eaten some of the foods mentioned above, or because your body is brewing something. As always: don't panic! If the fragrance of your pee seems strange, monitor it for a while (yes, I do mean you should take a calm whiff of your used toilet paper) and if your urine still smells suspicious after three days, make an appointment with a trusted doctor.

YOUR URINE SMELLS ...

Sweet: If your urine smells sweet or like fruit, it can point to a problem in how you metabolise sugar. There is always some sugar in urine — this is totally normal. If the smell is strong though, it could be because of diabetes.

Fishy: Does something smell fishy or foul after you pee? This can indicate a urinary tract infection. E. coli bacteria are usually responsible for this: they climb up into the urethra, where they cause irritation and ultimately inflammation. They also begin to break urine down while it's still in the body, hence the strong smell. Even in the absence of constant trips to the toilet and a burning sensation when you pass water, you could have a UTI. The infection could also have built up in the genital area rather than the bladder, which would also explain the fishy smell.

Like ammonia: Does a pungent smell of ammonia waft up to meet you from the toilet bowl? Is your urine also really dark? Then you've probably just had too little to drink and your body is a bit dehydrated. If you drink less than 1.5 litres of fluids a day, the urine is more concentrated and is no longer 95 per cent water. In this case, you of course need to top up your liquid levels and quickly.

How Pregnancy Tests Work and Other Things Urine Can Show Us

The stripe in the second window is going red — you're pregnant! One quick pee, and a few minutes later a pregnancy test can reveal whether a new life is growing inside you or not. But how does it actually work?

Pregnancy can be detected through the presence of the pregnancy hormone Beta-HCG (beta human chorionic gonadotropin) in the blood or urine. This hormone is produced by the placenta and ensures the production of progesterone, which in turn sustains the developing embryo. Beta-HCG can be detected in our bodies on the eighth day after sex and possible conception, so it doesn't help to panic and do a test straight after the condom has broken. Yes, there are special early tests, but the results are not as reliable.

By the way, pregnancy and other urine tests are most reliable when using morning urine, because it is more concentrated and comparatively free of residue from food or drinks. It's also best to collect the midstream as it's the cleanest (for advice on how to do this, see page 49).

A possible pregnancy is far from the only thing that can be read from urine. There are different kinds of tests for different conditions.

From the comfort of your own home, for example, you can find out how your urine is constituted using a *urine test strip*. To do this, you dip a strip of paper that has little colour fields on it into your urine.

The test strip changes colour, and then you can compare it with a colour chart to determine various indicators. For example, it can reveal the pH value of your urine, which influences your susceptibility to UTIs and kidney stones.

As well as a rapid urine test, your doctor can do a *urinalysis*, in which the most important data are put through their paces. To do this, you pee into a little cup and then your urine is examined in a laboratory. A urinalysis can be done to detect possible diseases of the kidneys or liver, certain blood diseases or diabetes, and even UTIs and urinary stones.

Anyone who is constantly fighting off bladder infections is sure to have done a *urine culture* with their doctor. Your doctor can use this to determine precisely what kind of bacteria is responsible for your bladder's misery. To do this, the urine is mixed with growth media for various pathogens and then checked to see which bacteria or fungi have made themselves at home. If your doctor finds bacteria through this test, she can also test which antibiotics will be most effective in fighting off your infection. An antibiogram is created for this, which shows the bacteria's resistances and sensitivities when exposed to various antibiotics.

The nerdiest of the urine tests is *24-hour urine collection*. As the name suggests, urine is collected for twenty-four hours and then examined in a lab. This is supposed to measure what exactly our body excretes. If, for example, we excrete too little of the waste product creatinine over the course of the day, it can mean that our kidneys aren't in such good shape and should be examined more closely.

Yellow, Golden, Transparent:
All the Colours of the Urine Rainbow

Show me your urine and I'll tell you about your health. This is because, along with its odour, urine's colour is also very revealing. And why is it usually yellow anyway? Why shouldn't it be pink or blue? Urine has what's known as urobilins (aka urochromes) to thank for its colouration. They are the metabolic products made when haemoglobin — a protein in your blood that carries oxygen around and makes it red — degrades, and they accrue in the kidneys. How intensely yellow your pee is depends on the concentration of urochromes in the urine: the more of these little metabolites there are swimming around in it, the yellower it is.

The colour can also change depending on what you've eaten. If, for example, you go through a phase where you're totally into carrots or beetroot, the colour of your urine will change to match your food: it will now leave the body *slightly reddish*. This is because these foods contain substances called carotenoids, such as beta-carotene, which dye the urine.

If you really feel like having a colour party on the toilet, you should eat lots of blueberries — they give urine a makeover with a slight *pink tone*.

Apart from that, we've already learned that urine gets darker when you haven't been drinking enough. If it's darker even though you have had enough to drink, something might be up with your liver or gall bladder. Have your doctor check this out immediately.

If, on the other hand, your urine is *very pale* or even

colourless, you've probably gone a bit overboard with the whole 'keeping hydrated' thing. You're most likely running to the toilet every ten minutes, too, so you'll know yourself what the problem is.

If your pee leaves your urinary tract looking *cloudy* instead of clear, it could indicate a UTI. *Foamy* isn't great either. If your urine is like this, get your kidneys examined.

And watch out: there's also *green* urine. This is a really rare phenomenon, but can happen when certain bacteria have spread or when you're taking particular medications. It's best to get it checked out.

Bladder Stones & Co.: Why the Bladder and Its Friends Don't Like Being Loaded Down

Diamonds are a girl's best friends — that is to say, rocks are a great look when they are worn around her neck, on her finger or in her ears. In the urinary tract, though, they're more like her worst enemy. Urinary stones are painful and can sometimes turn into something really dangerous. According to one study, one in eleven people will have a kidney stone in their lifetime, and this figure seems to be increasing: the prevalence in 1970 in the US population was only 3.8 per cent, but had risen to 8.8 per cent by 2010.[2] Meanwhile, Down Under, as many as 10 per cent of Australians are affected by urinary stones.[3]

Urinary stones are crystal deposits that form in the kidneys or bladder. This happens when the level of salts in the urine is too high and they crystallise. Once a salt crystal has formed, more and more layers are deposited on it, making it bigger. What was once a small, sweet little pebble quickly turns into quite the tough nut. These stones can be located throughout the entire urinary tract, where they lead to various problems.

Kidney stones, for example, develop directly in the kidneys. If, however, these stones go wandering and journey through the ureters, they change their name and become ureteric calculi. When crystals form in the bladder itself, they are called bladder stones.

If the stones play nice, *i.e.* stay small and manageable, they usually don't cause any problems. They swim around in your urine, and at some point are simply excreted along with it. If you're lucky, you won't even notice them.

It's only when urinary stones reach a certain size, attach to the bladder wall, or block the urethra that you slowly but surely become aware of your lodgers. You'll suddenly start to experience severe lower abdominal pain, micturition will be painful and no longer operate smoothly, and, from time to time, there can even be blood in your pee.

If you have the feeling a stone has made itself comfortable in your urinary tract, you should definitely seek medical attention. Your doctor will do a urine test to look for crystals, bacteria or blood. The urinary stones can then be examined more closely via an ultrasound, a CT or a bladder examination. If your doctor does indeed find a few rocks, there's no need to panic. If the urinary stones are small enough, they will vanish all by themselves — eventually you'll just pee them away. Certain medications can also be used to help wash them out. Larger urinary stones can be reduced to smaller pieces and removed during a bladder examination, that is, endoscopically. Usually, full surgical operations involving cutting and then sewing you back up again are unnecessary for urinary stones, and are only considered if there are complications. That sure takes a load off the urinary tract!

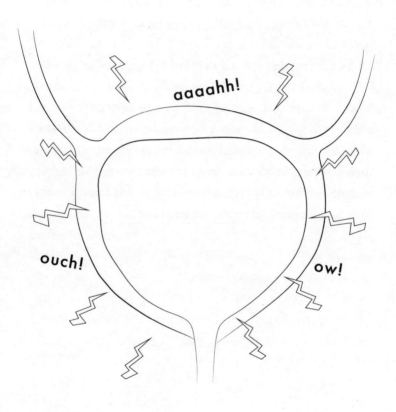

3.

Annoying and Painful: Urinary Tract Infections

Your tummy hurts, it feels like someone is drawing barbed wire through your urinary tract when you pee, and you have to run to the toilet every ten minutes. Know the feeling? Welcome to the UTI club! The nasty urinary tract infection that doctors like to call by the sexy name 'cystitis' happens when germs take up residence in the urethra or bladder mucosa, get a foothold and then multiply. Every second woman will have a bladder infection at least once in her life, which makes it the most common infection amongst women. Wow, congratulations — NOT! (For anyone interested: infections of the upper respiratory tract occupy first place when the entire population is considered.)

How You Get UTIs

The main reason we women are stricken with UTIs so often is that the female urethra, at just three to four centimetres, is pretty short. This means that bacteria find it easy to shoulder their way in and latch onto the bladder wall where they spark inflammation and irritation. The worst bacterial miscreant — sorry, it's a bit gross — is the gut bacteria Escherichia coli (that E. coli I mentioned earlier). According to some sources, about 90 per cent of UTIs are due to this bacterium that is at home in your rectum, aka the butt.[4]

Rest assured: the fact that gut dwellers sometimes lose their bearings and suddenly end up in the bladder has nothing to do with some kind of anatomical abnormality, and it can happen even if your hygiene practices are perfect. The anus, vagina and urethra are simply very close to each other in us women, making the entrance to the urethra no more than a hop, skip and a jump away for bacteria and germs. No wonder they so often take advantage of this tiny distance and show up in our bladder, totally uninvited. And, once there, they won't leave of their own accord. What rude guests! What's even worse is that approximately one in every four women who has had a UTI has it recurrently.

Every now and then, though, we do 'invite' them in ourselves. Not with greetings cards and pretty invitations of course, but with the wrong 'wiping technique' on the toilet. Are you a front-to-back wiper? Or do you perhaps prefer doing it from back to front? You should absolutely give up the latter habit. If you start at the back and then barge through

your entire genital area with the toilet paper, you're hauling germs from your anus across the vagina directly to your urethra. It's basically as if you're picking up the gut bacteria like hitchhikers on a highway and then dropping them off right in front of your urethra. Great service for the bacteria, terrible for your bladder! And so, my dears, let this really sink in: from now on, wipe from front to back!

Honeymoon Cystitis: Why Sex Often Heralds a UTI

Sex is, of course, also an important factor in the game of bladder infections. If UTIs often show up after your lover has left your bed (and your body), you most likely suffer from the legendary honeymoon cystitis. It's usually triggered during sex itself, because this kind of intimacy demands quite a lot of the vaginal area. A penis moving around the vagina, or contact with various other things like fingers, makes it easier for bacteria to find their way into the urethra. Since the anus and the vagina are right next to each other and since — in an ideal scenario — sex literally gets you quite moist, bacteria and germs can more easily sashay into your bladder.

Sometimes our partner's bacteria can also make us sick. I don't have to tell you here that we should wash our hands before making love, do I? Well, wash your paws before you handle your bae, and make sure that they do, too. It's not like he or she hasn't patted a dog, touched the railings at the train station, or held some loose change earlier in the day, right? Hands can quickly become potential germ slingers, and, along with cystitis, can also bring you presents like fungus or other infections.

But I don't just mean the stuff that might be on their hands. The genitals themselves can also carry disease. If, for example, your boyfriend has chronic prostatitis (inflammation of the prostate, which he probably wouldn't have even noticed), it could be these germs' fault that you keep getting UTIs. If you suspect this might be the case, your partner should definitely present himself at a urological practice. (There's more on this on page 81.)

To reduce the risk of honeymoon cystitis, it's always helpful to go to the toilet immediately after sex. This way, you flush the newly acquired germs straight back out of your plumbing. Not exactly romantic, I know, but it really does help.

There's also something that you could bear in mind during intercourse to minimise the risk of developing a UTI. It's to do with the position — or, to be precise, the angle at which the penis or finger penetrates the vagina. If you are particularly susceptible to bladder infections, you should choose a position where your partner doesn't move your urethra around with too much extreme motion and rubbing. Through this kind of friction, the urethra can swell up, making the path for bacteria even clearer. One approach that can work wonders is to experiment with different angles and shallower penetration. This has advantages beyond the prevention of inflammation: it can mean your partner will automatically massage your clitoris (outside and directly in front of the vagina) and the G-spot (on the upper wall right at the front).

And while we're on the topic: if anal sex is part of your repertoire, you should definitely make sure that the finger, penis, or whatever else found its way into your behind is thoroughly cleaned before it's used again elsewhere. It's similar to wiping back to front: you don't want to give gut bacteria a free ride in the direction of the urethra.

Contraception can also play a role here, especially if you are part of the diaphragm contingent. According to one study, women who use diaphragms get UTIs twice as frequently.[5] This is partly because the diaphragm can increase the presence of bacteria in the vagina, particularly E. coli. Diaphragms also cause bacteria in the vaginal area to thrive: the area becomes less acidic, making it the perfect breeding ground. The position of the diaphragm also means that it presses on the urethra, which can irritate it and make it more susceptible to bacterial infestation. So, it might be better to change your method of contraception if you often battle bladder infections.

Sex-Related UTIs:
When Your Bacteria Just Don't Get Along

If you often suffer from a UTI after lovemaking, it could also be because your and your partner's bacteria don't find each other as sexy as you and your boyfriend or girlfriend do. Every person has different bacteria on and in their body — this is totally normal. So when your bacterial tribes come into contact with your partner's during sex, they have to get used to each other. Most of the time, these foreign bacteria confuse and upset our vaginal flora and the bacteria that live there. Unfortunately, this can foster an infection in the long run. But don't worry, this doesn't mean you have to break up with your new sweetheart. Far from it. So that your bacteria can get to know each other better, you should just sleep with each other more often from now on. Yes, really! This way you give your bacteria the chance to make better friends. It won't happen overnight of course. But it's worth it. The more your bacteria like each other, the less cystitis will come to visit.

Baby, It's Cold Outside:
UTIs from Getting Chilled?

'Dress nice and warm, cover your kidneys and don't sit on a cold floor or you'll get an infection ...' Our mothers weren't exactly wrong — but they weren't entirely right, either. It's true that we get sick more often in the colder seasons than in summer. This isn't directly to do with the cold, though, but rather what it does to our bodies. To protect the immune system and keep our body temperature at a constant 36.8 degrees Celsius, the body really has to work to the max when it's cold. If it doesn't manage to, the organs aren't supplied with enough blood and can no longer work at full capacity. The result? The metabolism is curbed, the blood vessels narrow, and our defences stop working as they should. For the bladder, this means that the protective mucous membrane isn't properly supplied with blood, so it can't effectively fight back against invading bacteria anymore.

So, it is indeed possible to get a UTI through exposure to cold weather, but only because our immune system has been weakened generally. For this reason, it's true: always dress nice and warm, make sure you eat a healthy diet, and change out of your wet bikini or swimsuit straight after bathing.

Why You Get UTIs More Often Than Your Friends

Do you ever wonder why you're battling cystitis yet again, while your best friend will only ever have it once or twice in her life, if at all? Some people are sadly just more prone to bladder infections than others. Why? It hasn't really been established yet, but we have a few clues.

A healthy bladder wall has defence mechanisms which mean it can directly eliminate bacteria itself, thus protecting the cells underneath from attack. Well, it seems that some people don't have such a thick and healthy protective layer, so bacteria find it easier to get established and cause trouble on some bladder walls than others.

Alternatively, your urine might be too sweet. To kill bacteria, it needs to be acidic and contain ingredients like amino acids, ammonia and lysozyme. All these are excellent bacteria assassins.

It's also possible that neither the bladder nor the urethra is the problem; the vaginal area is. Normally, what's known as lactobacilli, aka lactic acid bacteria, live on and in our vagina. They render invading bacteria harmless right at the entrance to the vagina, and keep the pH value of our vaginal flora balanced. If there aren't enough of these good bacilli in the vagina, other bacteria can more easily get to the bladder. Antibiotics, for example, can be responsible for a decline in good bacteria. There are a range of probiotics that you can take orally or insert as suppositories which aim to replenish your vagina's good bacteria; ask your doctor for advice on these.

A healthy vagina is also furnished with a mucous protective layer, which, amongst other things, prevents bacteria from sticking to it and multiplying. If this mucous layer is either minimal or missing altogether, bacteria naturally have it easy.

Bye-Bye, UTI

At the slightest twinge or discomfort of the bladder, there's one thing that's really important: drink, drink, and drink even more. Still water and herbal tea are best. Bladder and kidney teas containing birch leaf or stinging nettle leaf are great at flushing things out. The more often you have to go to the toilet, the harder it is for the bacteria to get established in your bladder. Instead, they are simply flushed out along with your urine, kind of like a purifying shower. Yes, I know, it's super uncomfortable when every trip to the toilet hurts, but it can't be helped — just grit your teeth, relax your sphincter muscles, and bear it. If you don't drink enough, the bacteria can take up residence on the bladder wall and surrounds, completely unimpeded.

If the pain is unbearable, you can of course reach for painkillers. Ibuprofen works best here, by the way. In a study conducted in 2015, it was found that 70 per cent of a group of patients given ibuprofen instead of the antibiotic fosfomycin to treat an uncomplicated UTI were symptom-free seven days later. By way of comparison: for the women who took the antibiotic, it was only about 10 per cent more, *i.e.* 82 per cent.[6] Warmth can also help relieve the pain — just like with period pain, it can help to put a water bottle or heat pad on your tummy, or to take a hot bath.

Try to take it easy and look after yourself. You are sick, and you need rest. And something else that's really important: don't panic! Usually, this nasty UTI will disappear of its own accord in about three days, and you can spare yourself antibiotics and a visit to the doctor. But if you have the feeling that you're getting a fever, or you notice pain in your back, a trip to the doctor is a must. These can indicate that the infection has become more serious, and is spreading to your kidneys.

Antibiotics: Yay or Nay?

Sure, antibiotics are quick and (usually) they're reliable. And when your abdomen feels as if Chucky, Freddy Krueger and Pennywise are all up to mischief in there, it's totally natural for there to be nothing you'd like to do more than hotfoot it to the doctor's to stop them with the right antibiotics. It's just that this isn't exactly healthy in the long run.

Antibiotics can destroy bacterial structures that pose a danger to the human body. This might seem great at first, but antibiotics have some favourite secondary targets, too: good intestinal bacteria and lactobacilli. And, without intestinal bacteria, we are more susceptible to autoimmune diseases, food intolerances and becoming overweight — our entire immune system is put off kilter. As already mentioned, bacteria can more easily advance into the bladder in the absence of lactobacilli, and yeast fungi also have an easy job of it. If you're unlucky, the whole thing will set off a vicious cycle that gives you your next bladder infection.

Along with destroying good bacteria, the use of antibiotics can also lead to resistance over time; that is, it can make them less effective, as the bacteria evolve to survive them. Over-consuming antibiotics can happen when we take them too often, but a lot of people don't realise that we often ingest antibiotics along with our meals. Every time you eat or drink animal products, there's a high probability that you are unwittingly 'taking' antibiotics, as many animals raised for human consumption are dosed with them as a matter of course. In 2015, the German environmental NGO Bund für Umwelt und Naturschutz Deutschland e.V. conducted random sampling and found antibiotics in a total of 88 per cent of turkey meat products bought in discount supermarkets.[7]

Meanwhile, antibiotic resistance is steadily increasing: bacteria are effectively becoming 'immuner' to the effects of antibiotics, and at some point they will no longer be able to help us. It would be silly to demonise antibiotics and stop using them altogether: they are still an important tool in fighting infections. If your bladder infection doesn't go away on its own after a few days, if the pain moves into your lower back, or if you get a fever, you won't be able to avoid antibiotic treatment. But it shouldn't always be the first step. If you do require antibiotics, listen carefully to your doctor's advice and be sure to complete the full course of antibiotics as prescribed.

Avoiding UTIs

You can never be 100 per cent safe from UTIs. There are, however, some things we can do to at least minimise the likelihood of getting one. First of all, look at your shower routine: avoid aggressive shower gels, soaps or intimate sprays — our vaginas do a good job of keeping clean all on their own. If you do use harsh products, over time you risk destroying the genital region's natural flora, which, as we just learned, are very important in the fight against bad bacteria, germs and fungi. Keep your lady garden healthy by using mild, pH-neutral products: those that keep the vagina at a healthy, acidic pH value of under 4.5 are suitable.

Another important thing: the right clothes. For example, wearing trousers that are too tight and cut into your crotch doesn't just look silly, it can also encourage cystitis or other unpleasant conditions. If a seam rubs and chafes at your crotch it can cause tissue irritation, which in turn makes it easier for bacteria to spread and advance into the bladder.

During your period, change your tampons and pads regularly, ideally each time you pee. You should also make sure that the period products you use don't contain any artificial additives that could irritate the mucous membranes. It's safest to menstruate into 100 per cent cotton. If you use a menstrual cup, remember to sterilise it regularly, using boiling water only — you don't want to use alcohol or any other chemicals to clean it.

And I know you probably don't want to hear it again, but having enough to drink really is the most important thing!

Make sure your bladder is always well flushed, so that bacteria don't have the chance to get established. More generally, you should also make sure you eat a healthy diet, don't linger in the cold for too long, and are vigilant about your health. The stronger your immune system, the harder it is for bacteria to attack.

Is Cranberry Really a Magic Bullet in the Fight Against UTIs?

If you're plagued by UTIs, you'll hear this all the time: 'What about cranberries, they're fantastic for the bladder …' But is there any truth to it?

First, a bit of background: this little red berry, less commonly known as a fenberry or moorberry, comes from North America and grows on a dwarf shrub that spreads low to the ground. There are all kinds of great active ingredients romping around in these little red berries: first and foremost, we have the vitamins C, B6, and K, and the provitamin A. They also contain calcium, potassium, magnesium, folate and sodium. These are all fantastic, super-important substances, there's no question about it. But the most important weapon that cranberries offer in the fight against bladder infections is their so-called proanthocyanidins. This hard-to-say (and spell) substance is said to prevent E. coli — remember: these baddies are to blame for about 90 per cent of UTIs — from colonising the bladder wall and setting off that nasty inflammation.

Experts have been debating how beneficial this effect really is for a long time now. Some claim that there is little use in filling yourself up with cranberry capsules, juice and so on, because the berries have a long commute through the gut, liver and co. before they reach their actual workplace, the bladder, by which time there isn't really a lot of their great ingredients left. Others, meanwhile, swear to the protective effect of cranberries, and recommend their regular consumption. If you choose to drink cranberry juice, it's important to find products that are 100 per cent fruit juice, as sugar and emulsifiers used to preserve some drinks may feed infection in the mucosae. Pure cranberry juice with no added sugar can be quite sour and bitter, so you may prefer to take it in tablet form.

It may indeed be helpful to consume cranberry to try holding inflammation at bay. But these red wonder berries cannot conjure away an acute UTI that is already raging.

By the way, there's something called mannose that's similar. It is a building block found in plant sugars, and its manufacturers also like to advertise it as a miracle cure for bladder health. You can find mannose tablets in any pharmacy. Studies suggest this kind of sugar binds to bacteria, thus preventing it from attaching to the bladder wall. As with cranberry products, these studies are unfortunately not diagnostically conclusive, so their value can't be confirmed or denied.

Simple and Complicated UTIs

Doctors divide urinary tract infections into two different types: the complicated and the simple. Guess which one causes the most problems? You've got it: the complicated.

If you find blood in your urine, please don't panic! This is really common with bladder infections, and it's not necessarily a sign that you have a complicated UTI.

Complicated UTIs involve not 'only' pain when peeing and a constant urge to urinate, they're also accompanied by fever, chills and lower back pain. That's because these symptoms often occur when the bacteria are not just causing trouble in the urinary bladder or urethra, they've also climbed up into the pelvis of the kidney. If this happens, you should visit a doctor immediately.

Complicated UTIs can't just be waited out or 'peed away' with lots of fluids; antibiotics will be needed. Luckily, most cases of UTIs in young, premenopausal women who are not pregnant are the uncomplicated kind.[8] With lots of water and herbal tea, this less harmful variety of UTI will usually vanish by itself after a few days. If you are unlucky enough to be in the minority here, you should have the cause investigated, which will require a few tests. Your doctor will usually first take your medical history, examine your urine and do a urine culture. Through this, they should be able to find out which bacteria are responsible for your misery and which antibiotics will put them back in their place.

THINGS TO KNOW ABOUT SIMPLE UTIs:

- They occur in the lower urinary tract
- They're not dangerous
- They usually disappear on their own after a few days
- You can treat them yourself by having plenty to drink and staying warm.

THINGS TO KNOW ABOUT COMPLICATED UTIs:

- Along with pain when passing water, they are accompanied by fever and lower back pain
- They must be examined and treated by a doctor
- They are treated with antibiotics
- If they're left untreated, they can lead to kidney infections.

Hello Again! UTIs on Repeat

One bout of cystitis has barely subsided when suddenly you start feeling tingling and burning down below all over again. What? Another UTI already? That's impossible! Unfortunately, though, it's not. If cystitis comes to visit more than three times a year, physicians call it recurrent or persistent cystitis (I like to call it a zombie infection). And, because I know so many women who are fighting a running battle against recurrent cystitis, I'd like to try bringing a little light into the darkness.

Normal bladder infections often make a comeback

because the illness wasn't treated properly and some bacteria remained on the bladder wall. You might also be dealing with resistant germs against which the antibiotics prescribed were ineffective, or you might not have actually kept taking the medication for long enough. Just because you stop noticing the symptoms and feel healthy again, it doesn't mean that you are. Please, you really do need to keep taking antibiotics for as long as your doctor has prescribed them.

Your bladder wall might also be damaged. Bacteria and germs find it easier to hide in little tears or holes and then strike again. Of course, it's also possible that new pathogens have crept into the bladder. In that case, it's best to investigate why your bladder can't defend itself properly.

Unfortunately, there is no real way to prevent recurrent cystitis. You should of course always have plenty to drink so that the majority of the bacteria can be washed away, but to protect yourself as well as possible, it helps to have a competent urologist on your side, someone you trust and who knows your medical history. She can, for example, prescribe you with a course of suppressive antibiotics, *i.e.* antibiotics that you take over a longer period of three to six months. This should destroy the bacteria that are hiding somewhere in the bladder wall for good. To minimise the side effects of this long course of antibiotics, you should get advice from your doctor about how you can restore the good bacteria that are also destroyed by the antibiotics (see page 68 for more on this).

What are the benefits of vaccinations for UTIs?

Since 2004, a vaccine against UTIs has been available in some countries in the form of an injection — and for all you scaredy cats, the inactive bacteria can also be taken orally as capsules. The vaccine is supposed to provide basic immunity by purposefully introducing several inactive E. coli strains into the body so that your system gets used to the intruders and isn't as vulnerable to them in future. Both methods are said to reduce recurrent UTIs by 50 to 60 per cent.[9] At least in Germany, though, health insurers do not usually pay for this treatment because comprehensive evidence of its efficacy is lacking.

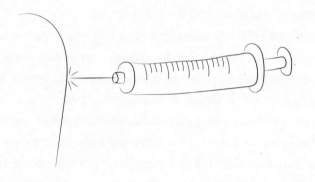

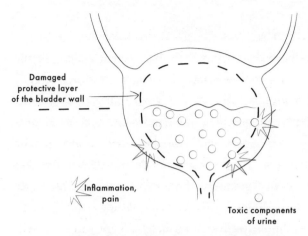

Damaged protective layer of the bladder wall

Inflammation, pain

Toxic components of urine

Constant Pain in the Bladder: The Dreaded Interstitial Cystitis

Inter-what? The word itself is confusing, which does not bode well. Translated into plainer English, interstitial means something like 'between' or 'gap-forming'. This is what it's all about: unlike the bladder infections now familiar to us, this one doesn't occur on the bladder wall but deeper in the tissue. Not enough is known yet about how this happens, but it is probably to do with what's known as the glycosaminoglycan layer. If this first, protective layer of the bladder is defective, urea can get directly to the bladder wall. This is not just extremely painful, it also causes inflammation to develop and become chronic.

Because there has not been a lot of research into interstitial cystitis yet, there are unfortunately few therapeutic outcomes or research results. This means that diagnosis usually happens through a typical process of elimination: are bladder stones to blame? Or fungi or a virus? Could genital herpes perhaps be the culprit after all?

After a cystoscopy and tissue samples, a precise diagnosis is finally achieved by distending the bladder and examining the way the bladder mucosa reacts. If you have interstitial cystitis, the mucosa will break and start to bleed quite profusely when the bladder is emptied, that is, when it shrinks to its normal dimensions. Yep, this is really horrible, which is exactly why this procedure is done under general anaesthesia. You will have no idea what's going on with this 'mucosal cracking' at all.

So, if you're diagnosed as officially suffering from this nasty bladder condition, your mucosa needs to get whipped back into shape. Doctors do this by introducing certain substances, such as hyaluronic acid or pentosan polysulfate sodium, directly into the bladder via a catheter. Pentosan polysulfate sodium is a substance manufactured from beech wood that resembles the protective layer of the bladder wall and supports its regeneration. There are now more and more medications with this agent in them, which, through controlled long-term use, are supposed to prevent toxic substances from getting through the bladder wall where they can cause inflammation and pain. Because the effect takes time, though, improvement can only be expected after about six months.

Another option is what's known as EMDA therapy (Electromotive Drug Administration). Here, a local anaesthetic and cortisol are administered to the bladder using a very weak electrical current, which is supposed to mean they permeate deeper into the bladder wall and have a more lasting effect. Antidepressants with pain-relieving side effects can also be prescribed.

Why Men Get UTIs Less Often Than Women

Instead of going on about their penises, perhaps men should boast about the length of their urethra from now on. After all, it is 20 to 25 centimetres long, making it up to seven times as long as us girls'. And because men's urethral opening and anus are farther apart than ours, men have to contend with fewer bacteria in their bladder, and ergo have bladder infections far less often. I'm sorry, but what I'm about to tell you is not fair: from a statistical perspective, only one in a hundred men will get a UTI in his life.

When a man does get a UTI, it's usually due to an enlarged prostate preventing his urine from being fully expelled. Over time, the urine left in the bladder accumulates bacteria and causes the infection. For this reason, men should visit a doctor as soon as they have even the slightest suspicion of a bladder infection. Rather than bladder infections, men are likely to get prostatitis, i.e. inflammation of the prostate gland. Before reaching the bladder, the bacteria that migrate through the urethra pass into the prostate and inflame it. This is often accompanied by fever, considerable pain, and a general feeling of sickness with aching limbs and chills. These symptoms appear because the prostate isn't a hollow organ like the bladder but a gland, so it has more of a connection to the circulatory system.

FEMALE URETHRA

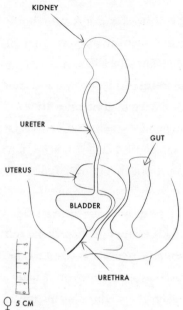

KIDNEY

URETER

UTERUS

BLADDER

URETHRA

♀ 5 CM

In comparison:
MALE URETHRA

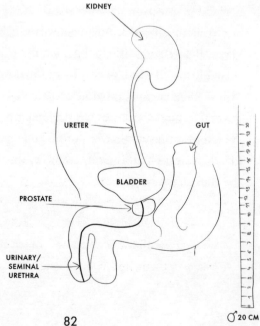

KIDNEY

URETER

GUT

BLADDER

PROSTATE

URINARY/
SEMINAL
URETHRA

♂ 20 CM

10 things I know about UTIs, aka things all frequent sufferers do

- Crop tops are totally out. You always ensure your kidneys and stomach are well insulated to keep them warm. High-waisted jeans? Excellent.

- Lying around cuddling after sex? Not with you. You rush straight to the toilet like you've been bitten by a tarantula. And beforehand you always down a big, sexy glass of water.

- Before getting down to business, you entreat your partner to wash their hands and penis thoroughly. Soap and towels are already handy …

- … or better yet, you do it in the bath or under the shower.

- A picnic outdoors or an open-air event? You're the square who always brings something warm to sit on. 'Better safe than sorry.'

- At work, your colleagues always ask you for ibuprofen or other painkillers. You're known as the wandering chemist.

- A hot water bottle in high summer? Sure, bladder infections don't take summer holidays and you have to calm your swollen abdomen down somehow.

- You know the shelf with the health products in your local drugstore back to front. Obviously, you're always on the lookout for new tablets and powders for bladder and urinary tract care.

- Orange, apple or perhaps even pineapple juice? Nope, all you ever have in your glass is cranberry juice. That's how you know all the brands, how much sugar they each have in them, and which is the best value for money.

Good etiquette when a bladder infection comes knocking

- Drink a lot to flush the pathogens out of your bladder (still water or herbal tea)
- Use heat and painkillers (ibuprofen) for the pain
- Rest and look after your own well-being
- If you have fever or pain in your lower back or sides, see a doctor immediately
- Take antibiotics for as long as your doctor has prescribed them

4.
Bladder Dysfunction: When Your Bladder Goes Crazy and You Can't Pee Properly Anymore

When everything in the bathroom is running smoothly, you don't really think much about peeing. But what about when you just can't go, however much you want to? Or the exact opposite: even when there's not much in it, your bladder reports an enormous, almost uncontrollable urge to pee roughly every twenty minutes. Quite clearly, the system is malfunctioning.

Overactive Bladder:
When Peeing Is a Full-Time Job

Normally, the bladder uses a feeling of pressure to politely request that you empty it at your earliest convenience. Once you've dutifully done this, it takes around two hours for it to fill up and send you to the bathroom again. If, however, your bladder starts demanding that it be emptied every half hour, or even more often than that, something has gone well and truly wrong. If this happens, you're probably walking around with an irritable — aka overactive — bladder. We're talking here about having to go to the loo more than four to six times a day. Which isn't actually that often, right?

If you've just mentally listed your daily trips to the toilet and realised with horror that it's more than five — please don't start panicking! Maybe you've just had more to drink today than usual, or polished off some extra diuretic foods like cucumber or asparagus. Perhaps your bladder is just a bit moody and oversensitive at the moment because you haven't been treating it very well lately (you've been sitting on cold floors, not drinking enough, or going wild with a new partner). Busted! Just observe your bladder more closely for a few days, keep it and yourself nice and warm and — most importantly — have plenty to drink. If it sends you to the toilet more than five times a day despite this TLC, it's time to go see a doctor.

Apart from increased frequency, another sign of having an irritable bladder is that when your bladder orders you to the toilet, she isn't as nice or polite as usual. No, an irritable bladder can rev from zero to one hundred in an instant.

Picture this: you're sitting in a café, drinking a flat white and listening eagerly to your friend's latest escapades, when suddenly a superhuman urge to urinate rolls over you. It hits you so powerfully that you can barely hold on. Waiting until the end of the story? Impossible. The urge to pee is so bad you feel like you're about to burst like a water balloon. You give in and sprint to the toilet, pull your pants down and just manage to reach the toilet bowl. And then you notice that your bladder wasn't even that full — there's much less urine than you thought there would be. Yes, ladies, you're probably dealing with an overactive bladder. If you don't manage to get to the toilet in time and wet yourself on the way there, then you're suffering from 'urge incontinence' (more on this on page 151). This can be a side effect of an overactive bladder (OAB), but it can also have other causes.

We're talking about OAB when:

- despite normal fluid intake, you have to go to the toilet so often that it disrupts your daily routine or stops you getting a good night's rest
- the desire to urinate comes on suddenly and is super urgent
- you might also leak a little because you don't manage to get to the toilet in time.

Pumped-up Bladder Muscles and Other Causes of OAB

If your bladder suddenly starts nagging and sending you to the toilet every twenty minutes, it's usually because of the detrusor, aka the bladder muscle. It is just too strong, so it contracts at full throttle even when it doesn't have much in it. Why is it acting like a bodybuilder and doing workouts at twenty-minute intervals? Well, no one really knows for sure.

Experts in the field assume changes in the bladder wall have made the sensors that measure pressure in it overly sensitive. This means they sound the alarm for immediate evacuation of the bladder much too early.

Apart from this, the neural pathways that carry the pressure signal from the bladder along the spinal cord to the brain could also be damaged. Damage to the spine from a prolapsed disc, for example, or due to neurological conditions such as Parkinson's or multiple sclerosis, can change how the nerves work, making the bladder muscle contract more often than it actually should. This is what urologists call a neurogenic bladder (more on this on pages 122–3).

If you have often been afflicted with UTIs, it could be that your bladder wall has simply been attacked by bacteria too many times, meaning that the stretch sensors don't work properly anymore.

Unfortunately, though, it's not at all uncommon for detrusor overactivity to be 'idiopathic', that is, 'due to unknown causes'. If this is the case, all doctors can do is alleviate the symptoms, not treat the cause.

But alongside an overactive detrusor, there are lots of other reasons for OAB. For example, there could be foreign objects in your bladder, such as urinary stones or even a tumour. Or you could have an infection. This is why it's so important to have it carefully investigated via an ultrasound or even a cystoscopy.

Your doctor should also check whether you might have a prolapsed bladder or uterus. This can happen when the ligaments and pelvic floor are loose or too weak to hold the bladder and uterus securely where they belong. This can cause the organs to slip down, which, along with other disorders, can trigger OAB and quite a strong urge to urinate (more on this on pages 144–7).

Lack of oestrogen can weaken the pelvic floor and connective tissues and then lead to bladder or uterine prolapse. If this happens to you, the hormones can be built back up again through suppositories that you insert into the vagina.

More generally, changes to the balance of your hormones can also foster an irritable bladder. If our oestrogen levels sink, our bladders might react more strongly to certain substances in our urine, thus sending us to the toilet more often; and less blood is supplied to the mucous membranes, which makes us more susceptible to inflammation.

If doctors have turned your urinary tract completely inside out, screened it and put it back together again, checking all organic causes and ruling them out along the way, only one possible culprit remains: your psyche. OAB is the most common gynaecological condition that cannot be fully attributed to physical causes.

Not every bladder-voiding disorder has an organic cause: they can sometimes be psychosomatic. So, the ancient Chinese proverb 'the bladder is the mirror of the soul', and German sayings such as 'the bladder cries out' or 'your tears took the wrong route' (*i.e.* they're coming out of your urethra instead of your eyes!) have a kernel of truth.

Many types of alternative medicine see the bladder as a passageway for our emotions, for example suggesting that if we don't vent our feelings, they may be expressed physically through bladder problems. Overactive bladder, which puts a lot of pressure on the body, can be read as a sign that we are putting too much pressure on ourselves and setting unrealistic goals. If no physical causes for OAB are found, it would be sensible to take a closer look at your psychological state.

Possible reasons for a constant urge to pee:

- The bladder muscle contracts too early
- You have a UTI or urinary stones
- You are suffering from a bladder or uterine prolapse
- Your hormone balance has gone haywire
- There are psychological reasons for your bladder problems.

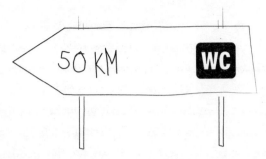

10 Things You Know If You Have to Go a Lot

This one goes out to all you frequent tinklers, tiny bladder owners and whizz kids — in short, anyone who needs to go to the toilet a lot.

Being ashamed of your body is so passé! That's why these ten facts are here, just for you: great people with great bladders that just happen to always crave attention.

- Better safe than sorry: you never fail to use the bathroom before leaving home, even if you don't really need to.
- If your bladder does start twingeing while you're out, you always know exactly where the best place to go is.
- The toilet cleaners know you by your name and are always nice to you.
- You always have enough loose change with you, in case you have to pay to pee …
- Coffee before a long drive or when exploring a new city? No way! You know which alternatives will pep you up.
- You're the running joke in your circle of friends, and it's a long time since anyone asked you how you can possibly need to go again.

- Despite this, everyone is always surprised at how quickly you can go to the toilet. Practice makes perfect!
- Your favourite seat in the cinema or on an aeroplane? Right on the aisle, of course.
- In the office, no one is surprised if you're not at your desk: '*She's gone to powder her nose again, she'll be back in 10 seconds.*'
- When your friends are complaining about uncomfortable gynaecological procedures, you just give a tired smile … by now, you're basically a urologist yourself.

Treating an Overactive Bladder

It's not at all easy to find out why some bladders suddenly start acting up, and it's just as hard to treat them properly: often, OAB can't be cured and all that can be done is alleviate the symptoms. To give you an idea of roughly what kinds of treatment options are available, here's a summary.

Medication for OAB

To at least alleviate the symptoms of overactive bladder, urologists will usually prescribe you a type of medication called an anticholinergic. An entire armada of these medications is available, and they all work more or less according to the same mechanism: the active ingredient attaches itself to what's known as muscarinic receptors, some of which are found on the bladder wall. By doing this, it prevents the messenger substance acetylcholine from being transferred from the nerve cells to the detrusor. The result: the bladder muscle doesn't get the message to contract as early and we don't need to go to the toilet as often.

It sounds great, but unfortunately this benefit comes with all kinds of unpleasant side effects. Muscarinic receptors are not just in the bladder, they're also in lots of other parts of the body. In the eyes, for example, and in the gastrointestinal tract and the salivary glands in the mouth. So, while your bladder might be less irritable, you can also get indigestion, mouth and skin dryness, and impaired vision.

The good news is that because there are lots of different medications on the market, you have a range of pills to choose from, and some may suit you better than others. And don't worry: it's totally normal for it to take a while for you to find the right medicine for your bladder.

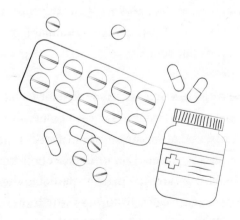

Bladder training: how to educate the bladder

Along with swallowing pills and waiting for the results, you can also do something more active for OAB: bladder exercises. Don't worry, your bladder doesn't have to start doing crunches or planking (although that would be kind of cute); instead you can do more subtle exercises to teach your bladder to calm down.

And how do you do something like that? Easy: stop going to the toilet as soon as your bladder wants you to and instead resist the urge for a bit. Tell your bladder something along the lines of: 'Sorry, it's not your turn yet. You'll have to be patient for a few minutes.'

If you want to give it a try, some tips for training your bladder are below. But first, a warning: because the bladder has complex nerves that we can't control, bladder training for OAB is controversial. In some circumstances, holding in your urine for a long time can even make the problems worse. So observe your symptoms closely during the training, and you may want to talk to a doctor if you decide to go ahead with it.

Here are a few tricks to keep your bladder tranquil. Start with trying to control the urge to go by sitting down. By sitting on a comfortable chair, you can press your pelvic floor down a bit. Bending your body into a bit of a swayback — i.e. raising your chest and arching your lower back — works really well. This way, you displace the pressure and have the feeling of being 'leakproof' down below.

It can also help to cross your legs, squeezing your thighs together and contracting your urethral sphincter. By doing this, you can dampen the urge to pee and delay going to the toilet for a good twenty minutes. Concentrate, and take deep breaths right into the abdomen. This will relax you and your bladder. The longer you avoid giving in to the urge and stay bravely seated instead of running to the toilet, the quicker you'll win back control over your bladder instead of letting its rhythm dominate you. Down with the bladder dictatorship!

It's easiest to work on delaying your trips to the toilet gradually. In other words, you don't have to hold out for two hours right from the get-go. This would be too much for you and your bladder at the start. Begin by delaying each trip to the toilet by five minutes. Once you've got used to this, increase the delay by ten minutes, and so on.

All you control freaks out there can also record your toilet times in a journal. Then your progress will be clear. It may seem like slow work, but over time you'll see that if you stick to the training regime, you'll regain control over your bladder and once more be able to decide for yourself when it's time to go to the toilet.

Botox, Muscle Relaxation, Electrical Currents: Alternative Treatment Options

Have you tried practically everything out there, but your bladder is still nagging you, or the side effects of the drugs are too much to handle? Then it's time to bring out the big guns.

Calm down: Botox for the bladder

You've probably heard of Botox, the nerve poison that people have injected into their face to make them seem younger, occasionally going overboard until they resemble a waxwork. But Botox really can do more than deface people — um, I mean, give them beautiful faces. Botox can be used for migraines and hyperhidrosis (excessive sweating). And it can also be used to immobilise an overactive bladder. Botox blocks the interface between nerves and muscles, so the muscles can no longer spasm or contract — which happen to be an irritable bladder's two favourite hobbies.

The procedure is as follows: to check exactly how your bladder is going and where the needles should be applied, your doctor will first do a cystoscopy. The Botox is injected

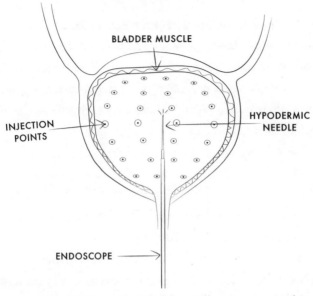

into several locations on the bladder wall using a special needle. It usually happens under general anaesthetic, so you don't notice the needle at all. You and your doctor will decide how many units of Botox should be injected before the procedure, but usually the amount is between 100 IU (international units) and 200 IU. If you inject too much at first, there is a danger that the bladder muscle will be made too weak and won't be able to contract at all anymore. If this happens, then instead of needing to run to the toilet super urgently, you will only be able to squeeze out a tiny amount of urine, if any at all (more on this on page 108). To avoid having residual urine, you would then have to catheterise yourself for a while, *i.e.* suck your urine out through a tube (more on this on pages 120–21). You don't want this to happen of course, and that's why doctors prefer to start by administering just a small amount of Botox, and then see how your bladder reacts.

If all goes according to plan, the Botox procedure takes about ten minutes. When you wake up from the anaesthesia, don't be shocked to find a catheter dangling out of you. And it will stay there until the next morning, too. After all, your bladder has been agitated quite brutally. It's earned a break from collecting and storing urine, and there's a danger that it might now become inflamed. After the catheter has been removed, you will have to check that you can still pee normally. What's the pressure like? Does everything trickle out happily, or is there residual urine? After voiding, you will have an ultrasound done on your bladder to check it, and if this is looking good you'll be able to go home.

If you have this procedure done, don't be surprised if the Botox doesn't kick in straight away and you still have to go to the toilet every twenty minutes for a while, or if you even wet yourself. The full effect of Botox usually only develops after about two weeks. And sadly, it doesn't last for ever, either; it works for about three to six months, depending on how quick your metabolism is. If you were happy with the result, though, you can always repeat the process.

Progressive muscle relaxation

How exactly are you supposed to relax properly and find peace when your bladder is butting in on you every twenty minutes? Well, you could actively relax your muscles in those very twenty minutes between two trips to the toilet, through progressive muscle relaxation. You can choose for yourself which muscle groups you do and how long you hold the tension, which of course also depends on how much time and leisure you have in which to practise relaxation.

Lie down on a comfortable surface, rest your arms gently beside you and then inhale and exhale. You will first tense specific muscle groups, and then consciously relax them. So, once your bladder is empty, you're chilled out on a comfortable surface, and you won't be disturbed for a little while, let's go:

- Let's start with the right hand. Ball it into a fist, hold it for a moment, inhale and exhale consciously, then release it. Feel it: inhale and exhale. Now let's do the same thing with our left hand, then with both hands at the same time.
- Now come the upper arms. We'll start with the right arm again. Press it consciously into the ground and then relax it again. And, you've guessed it: we'll do the same thing with our left arm before tensing and then relaxing both arms at the same time.
- Now the stomach area: to consciously tense it, just pull it in tightly while at the same time sticking the chest out. Hold the tension, inhale into the chest and then relax.
- Next we'll press the legs into the ground as firmly as possible. Let's start with the thighs, tensing and relaxing them, and then work like this through the calves and down to the heels.

Of course, you can add other muscle groups into your exercises. You can't really go wrong with it. Naturally, you can't expect any miracles immediately, but in general, it makes the body feel better. By practising progressive muscle relaxation regularly, you can become more attuned to the different muscle groups and become generally more relaxed, which can also transfer to your bladder and its activity.

Insert an electrical current here, please!

Electrical stimulation might sound like a nasty torture method, but it can be a real blessing for anyone plagued with OAB or incontinence. And it doesn't hurt at all, either. In electrical muscle stimulation, or EMS for short, you insert a probe into the vagina that emits low-frequency electrical pulses at short intervals. It is supposed to stimulate the fibres of the motor nerves that control the muscle contraction, which should in turn re-establish the balance between muscular inhibition and activation.

If your urologist deems electrotherapy appropriate for you, she will explain which device to use, which power setting is most appropriate, how to insert it, and everything else you need to know. It sounds pretty out-there, but don't be scared; it doesn't hurt, it just tickles a little. I promise!

Pacemakers for the bladder, aka sacral neuromodulation

Do you suspect that your bladder muscle is so steely that even a high dose of Botox won't be able to rectify it? Then perhaps you need a 'bladder pacemaker'. Yes, there really is such a thing. This pacemaker is implanted under the skin on the sacrum (the bony spot just above your butt) and is connected via an electrode to the sacral nerve. These nerves conduct the urge to pee along the spine and up to the brain. The pacemaker uses electrical impulses to stimulate these nerves so that they work normally again and send the brain the right messages.

Of course, therapy is only an option if your pee problem is due to defective nerves. To find out whether this kind of external nerve stimulation is an option for your bladder, you'll use a trial stimulator for several days to see whether and how your nerves react. A thin wire with electrodes on it is inserted through the skin and placed above the sacrum, which can then be activated by means of a pulse generator. This should trigger the sensors on the bladder wall, which should spur on your bladder muscle again. If the test phase goes well, a pacemaker can be implanted and help you urinate for years.

Nocturia: The Nightmare of Constantly Needing to Go at Night

You've just whispered 'Jack, I'm flying,' as you soar over the city on your white unicorn, when something unwelcome tears you away from the wonderful dream: a full bladder. Annoyed, you peel yourself out of your cosy blanket and hoof it to the toilet, your eyes puffy and half-closed. Back in bed, it doesn't even take two hours for your bladder to sound the alarm again. (At least this time you were being chased by a crazed robotic chicken.) You never actually need to set an alarm to wake up in the morning, either: every day, your bladder forces you out of bed half an hour earlier than you need to get up.

If you feel like this is you, then you're suffering from what's known as nocturia. This is when the bladder has no closing time and wants to be emptied more than twice a night.

This isn't just annoying at night; it weighs you down during the day, too. Being repeatedly ripped out of the slow-wave sleep phase makes it hard to fall asleep again, so you end up knackered to the point that you have difficulty concentrating, and your intellectual functioning can even be impaired. According to Danish researchers, nocturnal bladder weakness can decrease productivity at work by as much as 24 per cent.[10] Researchers have calculated that nocturia causes economic losses running into the billions: according to a study by the RAND Corporation, the US GDP could be approximately 44.4 billion dollars higher if there were no nocturia and its citizens could sleep through the night in peace.[11] Beyond this, nocturia can also lead to depression in the long run: sleep deprivation can make you deeply unhappy. But what can you do about it?

First you need to find out why your bladder is terrorising you at night. Urine production could be to blame. Your kidneys might be busy bees that don't know when to knock off, and therefore keep producing urine all night. Or the problem could be in the bladder itself. Normally, we expel about 80 per cent of our urine during the day. For some people, though, it's only 50 or 60 per cent, meaning that what's left over needs to be voided at night. And if you consume lots of fluids in the evening, it's obvious that you'll have to get up more often during the night. In theory, you could plan it all in advance: if you drank one litre before going to sleep, and assuming that your bladder has room for about 500 ml, you'd need to get up twice. If you had 1.5 litres, you'd have to make three trips, and so on and so forth.

To help you sleep peacefully, try having less to drink before going to bed; if, however, you suspect that you do indeed have nocturia, you should definitely consult a specialist.

Out and About and Your Bladder Niggles: Suppressing the Urge to Pee

It's the nightmare of every whizz kid, every pea-sized bladder owner, and all you ladies who constantly have to powder your noses. You've just gone out, there's not a toilet to be found and your bladder is calling. So now what will you do? Pee your pants? Cry? Both at the same time? Luckily, there are a few tricks to placate an oppressive bladder on the go.

Sit down

See above for controlling OAB. If you sit down when you have a strong urge to urinate, you can inconspicuously press your pelvic floor into the seat, for example, by arching your back a little or crossing your legs and pressing your thighs together. It also works psychologically. You have the feeling that you're leakproof because the chair is supporting you. You can delay going to the toilet for a good quarter of an hour this way. So, if you're out and your bladder starts to squeeze, quickly find the next park bench or another public seat. Sit down for a little while, stretch, arch your back, and then onwards.

Press on your clitoris, aka 'tie your shoelaces'

If your bladder reports for duty when you're out, pressing on the clitoris or its surrounds can help (men should put pressure on the head of the penis). And why, you might ask? In short, it leads to the tension in the bladder wall decreasing and the tension in the sphincter increasing. As a result, the urge to urinate becomes weaker. If you feel uncomfortable pressing your hand to your genitals in public (fair enough), you can use the shoelace or stone-in-your-shoe trick: kneel down and press against your clitoris with your heel. This can be quite awkward when you're in the middle of a busy footpath or find yourself somewhere else that's overrun with people; in that case, squeeze your lips together and seek out a quiet spot.

Breathe properly

Breathe. Inhale deep into the lower belly. Like in meditation, it helps to inhale deeply, count to three, exhale, and count to three again. This technique has three advantages: first, your mind is kept busy concentrating on breathing and counting properly. This calm, deep breathing also soothes the body and the soul, so all the pressure is eased for a little while. And, last but not least, the deep inhalations and exhalations give you the feeling that your bladder has more space, so it doesn't have to contract so small. Try visualising it, and perhaps you can suspend the pressure that way.

Distract yourself

This is very hard when your bladder is putting crazy pressure on you. But try to distract yourself anyway and think about something else. Like concentrating on your breathing, you could, for example, sing a song in your head and try to remember all the words. Your bladder's current favourite song is probably 'Under Pressure' by Queen (have fun getting that out of your head). Reciting a poem is also great. Or ask yourself what the person walking in front of you ate for lunch. Just try to steer your thoughts away from your bladder.

Suck on an imaginary sweet

Does this sound like nonsense? Let me explain: according to some alternative health practitioners, the incisors in the upper and lower jaw are connected to what's known as the urogenital system, which would make them relevant to the bladder, kidneys and genitals. Now, when you stimulate these oral reflex zones by running your tongue from the back and middle of the lower incisors over the back and middle of the upper incisors, it can affect your pee-quipment: it lessens the urge to urinate. In general, it may be a good idea to pay attention to what's going on in this area of your mouth. Have you perhaps had inflammation, implants or fillings in your front teeth? This could have a negative impact on your bladder and be the cause of any possible problems.

Roll back and forth on the balls of your feet

This is exactly why small children can never stand still and instead trip from little foot to little foot when they really need to go. The quick changes of pressure on the balls of the feet tighten the pelvic floor muscles, lessening the urge to urinate. What's known as the kidney–bladder meridian in Ayurvedic teachings also lies in the ball of the foot. The bladder muscle is inhibited by this seesawing on the sole of the foot, suspending the urge for a little while.

Talk to your bladder about its capacity

'Come on now, bladder, we can last another five minutes. Then we'll be at the next toilet …' This or something similar is what you could say to your bladder when it is terrorising you and needs to be soothed when you're out. Have the conversation in your head of course, not out loud! During this little 'performance review', picture the urinary tract and remind the bladder of its elasticity and size! It's important that you always formulate everything positively and don't focus on the problem that you need to go soooooo badly that you might be about to wet yourself. It might sound odd, but talking to yourself like this can increase your feelings of security and control.

Lucky last: why it's really important to keep your body warm

In the chapter on UTIs (page 61), we already learned about the importance of keeping your body, and especially your kidneys, warm. But did you know that getting chilly can also make us need to go to the toilet more often and more urgently? This phenomenon is called cold diuresis. Our blood vessels constrict in the cold to minimise the loss of heat, resulting in increased blood flow in the centre of the body and better blood supply to the kidneys. And when our kidneys have increased blood circulation, they inevitably produce more urine. So wrap up nice and warm before leaving the house when it's chilly!

Contracted Bladder: Can Your Bladder Really Shrink?

As we have learned, a healthy bladder has a capacity of 350 to 550 ml in us ladies. But it is also possible for this to decrease. Yes, the bladder can actually shrink such that it can store less than 100 millilitres, or in some cases just 15 per cent of its standard volume. Specialists call this a 'contracted bladder'.

The symptoms, such as frequent and urgent tinkling, are very similar to overactive bladder. The difference is that when you experience OAB, the bladder is a normal size, but it can no longer be filled to the top. In the case of a contracted bladder, on the other hand, it can be filled to the brim, but its storage capacity is much lower.

A bladder infection that hasn't been cured is often responsible for this misery. If it isn't treated properly and never fully heals, the bladder wall can harden; the tough, inflexible bladder will then no longer be able to expand to its proper size.

If it hasn't been caused by an unhealed infection, a contracted bladder could also be a long-term consequence of the dreaded interstitial cystitis. Otherwise, it could be down to damage to the bladder wall, *e.g.* from an operation near the bladder that has scarred over.

By the way, the fear that you might yourself be responsible for your bladder shrinking because you've peed too often is unfounded. The bladder cannot become anatomically smaller from contracting too frequently. As we have learned, what can happen instead is that trigger-happy pee-ers eventually become accustomed to higher frequency and are hyper-responsive to the urge to pee. There's more on this on pages 46–7.

Doctors can only help you once they know why your bladder has shrunk. Some of the medications used for overactive bladder can also be used here. They don't enlarge the contracted bladder, they just alleviate the symptoms and help you to normalise your urination, thus minimising trips to the toilet.

In extreme cases, an operation might be considered. Through what's known as bladder augmentation, the bladder is artificially enlarged to allow more urine to fit in again. If this isn't possible, doctors can even install an artificial bladder for you.

Battening Down the Hatches: Why You Can't Empty Your Bladder Properly

You need to go to the toilet super urgently and you feel like you're about to have an accident and wet yourself. Once you're on the toilet, your urine will gush into the bowl like the Niagara Falls, giving you an incomparable feeling of relief. Or at least, this is what you imagine when you're on the way to the WC.

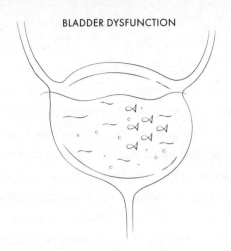

Once you're there, though, what happens? Nothing. A pathetic trickle of three or four droplets lands in the bowl. Strange!

Something must have gone wrong in the bladder's dispatch department (to read about how this *should* work, see pages 16–18). Someone who suffers from a voiding disorder either cannot get urine out of their bladder at all, or only by squeezing really hard. And, although their urine won't stream into the toilet, those affected often complain of annoying leaks: they've barely finished their business and pulled up their pants when a few drops of urine trickle out.

There are various reasons why your bladder might suddenly go haywire, so it can take a while for those affected to find out what's bothering their bladder.

Symptoms of a bladder voiding dysfunction

- Difficulty passing water
- A weak or broken urinary stream
- Taking a long time for the initial stream to start
- Having the feeling of not being completely 'empty'
- Sometimes leaking just after you've finished

The perfect urine stream

What does the perfect stream of urine actually look like? We all pee differently, of course, so every stream of urine is individual, too. The strength of your stream depends on various factors. One is the pressure, that is, how hard the bladder muscle goes for it when it's pressing out your pee. Obviously, in order for this pressure to be released properly, it's also important that the exit route is clear, so the diameter of the urethra needs to be right; if it's too narrow, it leads to traffic jams and it can then take longer to urinate.

A healthy, well-running stream of urine should flow at a rate of 15 ml per second. You can find out what your flow rate is through what's known as a uroflowmetry test: you pee into a funnel that collects the urine, and a device measures the exact volume of urine per unit of time, which is then represented on a line graph that your doctor can evaluate.

I don't want to sound superficial, but, along with pee speed, looks are also important when it comes to the perfect urine flow. It should be consistent, and as straight as possible. If your stream reminds you of your neighbour's garden sprinkler and you have to clean your toilet thoroughly after peeing, it's definitely something to monitor. If your urine doesn't flow straight into the toilet, instead splitting into two or even fanning out, your urethra or its orifice might have narrowed.

Besides all this, the urine should also pass in one continuous stream. What does this mean? Your pee should not be piddled out in small, jerky little portions. If it sounds like you're sending out messages in Morse code when you go to the toilet, something is not right with your bladder's drainage system. It might be because the urethral sphincter is opening and closing of its own accord while you pee, or a urinary stone might be cheekily blocking the bladder's exit.

Why you shouldn't push

If your urine stream is weak or uneven, it's tempting to start trying to help by pushing. This is not a good idea: only very few people manage to push and simultaneously relax the pelvic floor; pushing usually results in pressure being exerted while the pelvic floor keeps the exit closed. This pressure is not good for your urinary equipment and, in the long run, can damage the bladder wall.

The dangers of residual urine

Often when you have a voiding dysfunction, you'll also have residual urine in the bladder. This lingering urine is not just annoying because it gives whoever is carrying it the feeling of always needing to go to the toilet, it can also be dangerous. You might be thinking, 'Why? The bladder is never completely empty anyway, a bit of urine always stays behind …' While this is true, it's usually only about 10 ml, which is so little that we don't even notice it. When we call it residual urine, we're talking more like 100 ml. And believe me, that's enough to notice.

The danger with residual urine is that it acts like stagnant water and can become a hotbed for bacteria. Because the bladder is not being completely flushed out on a regular basis, germs find it easier to multiply and settle on the bladder wall. This can result in horrible inflammation and infections. The bacteria could even climb through the ureters into the kidneys and cause kidney damage. Too much residual urine in the bladder can also develop into a big urinary tailback into the ureters and kidneys, which can lead to kidney disease and even uraemia (when your blood has high levels of toxins such as urea — gross). On top of all this, the probability of developing urinary stones also increases, because the stagnant urine gives the little stones all the time in the world to grow by building layer upon layer of salts.

If you're not sure whether you've got residual urine hanging around, just do the following: after going to the toilet, press your hands into your stomach in the area where the bladder is. If this makes you want to pee, there's still urine in your bladder that shouldn't be there.

The bladder muscle is too weak

What the bladder muscle does to the extreme with irritable bladder, it does too little of with voiding dysfunctions: contracting and pressing urine out evenly. There can be several reasons why it might suddenly get lazy and not feel like doing its work properly anymore. One of them is age. As you grow older, the muscle may just become too weak, so it's no longer capable of contracting properly to press out the urine. If you're still a spring chicken, though, your nerves could be responsible for your misery. They might just fail to communicate to the bladder muscle that it should contract, so the muscle itself can't do anything to help. Perhaps it would in fact like to go to work, and is waiting longingly for the go-ahead from above. Nerve diseases, such as Parkinson's or multiple sclerosis, could be responsible for this, as of course can injuries such as a slipped disc or spinal stenosis. So, if no causes can be found through examination of the bladder, your doctors should definitely take a closer look at your spine and, if necessary, refer you to a neurologist.

Another thing that can happen is that the bladder can be overstretched because you empty it too seldom. It is full to the brim too often and for too long, and the sensors and the muscle stop working properly — they just don't know what to do anymore (see pages 45–6 for more on this).

Finally, as with irritable bladder where the overactivity of your bladder muscle is sometimes idiopathic, it might not be possible to find any medical cause for the problem.

The urethra is too narrow

Do you ever wonder why all you get is a puny trickle of pee dribbling into the toilet and it takes you forever to empty your bladder, while in the toilet next to you, a colossal waterfall is gushing down? Well, perhaps your urethra is too narrow. You might say: 'If the exit starts narrowing, peeing becomes harrowing.' A urethra that is too narrow can cause the flow of urine to change, so that instead of being peed out straight, it is

URETHRA

STOP

divided or even fanned out, and looks more like an elaborate canal system than a rushing river.

When it's hard to empty their bladder, those affected sometimes try to help it by pushing hard with their stomach, which — as we've learned — can do more harm than good. Apart from this, residual urine is often left behind, which can in turn lead to bladder infections.

If you are diagnosed with what is called a urethral stricture, treatment can include surgery to widen the narrowed canal by stretching or cutting it.

Urinary stones are stuffing up your bladder

If a bladder stone makes itself comfortable right in front of your bladder's exit, it acts like the plug in a basin. This obviously causes problems expelling your urine normally. The good news? Urinary stones are really easy to get rid of — see pages 57–9.

Your pelvic floor is uptight and isn't working properly

As we've learned, one of the reasons the pelvic floor is so important is that it holds our organs in place and seals the bladder off with the external sphincter. So, you would think it'd be good if everything in your pelvis was always nice and taut, firm and super powerful, right? Nope, it's not that simple. First of all, it's quite painful when your pelvic floor clenches up too much. It pushes, pinches or pulls. On top of this, it can lead to the various muscles working against rather than with each other. The technical term for this is DSD (detrusor sphincter dyssynergia): in other words, malfunction of the pelvic floor. Along with other annoying symptoms such as vaginismus, that is, a tense vagina and pain during sex, it can also prevent proper micturition. The bladder muscle strains because it wants to press out urine, and the internal sphincter muscle does the same thing because it wants to keep it in. Obviously, it's not straightforward for the bladder to empty itself in these circumstances.

Hinman syndrome, or non-neurogenic neurogenic bladder

At first, everything seems to point to the nerves being responsible for your bladder drama. When they're checked, though, it emerges that they're doing their job diligently on the micturition front. With 'non-neurogenic neurogenic bladder', aka Hinman syndrome, the problem is with the bladder's acquired behaviour. Sufferers started holding their urine in for too long in early childhood. At some point, the bladder became accustomed to this behaviour, and now it can't help but store too much urine.

Children are often affected by this kind of voiding dysfunction because they don't yet know how to go to the toilet properly. If it is not recognised, this incorrectly learned peeing behaviour can continue into adulthood. As is so often the case, the earlier the problem is detected and treated, the easier and more promising the chances of recovery.

If those affected stop being able to expel their urine at all, they are at risk of acute urinary retention. This is not just really uncomfortable, it's actually a urological emergency.

Urinary retention and why it's dangerous

Urinary retention is when the bladder is full to bursting but you can't empty it yourself at all. If this happens gradually, apropos of nothing, and without you really noticing it, it can develop into bladder overflow. Here, the bladder simply starts letting the urine out in small drops. Even without a toilet.

Urinary retention can also come on suddenly. In both cases, the condition is not just inconvenient and painful, it can become quite serious. That's because when our bladders are full, they are much more sensitive and susceptible to injury than when empty, so it's easier for the bladder to rupture (more on this on page 41). So if urinary retention does come on suddenly, urologists will insert a catheter to quickly drain urine out of the bladder, thus relieving the pressure.

Depending on why your bladder has gone on strike, different treatment options are available. Is the problem with the bladder itself, for example, and the bladder muscle is too weak? Or is the sphincter refusing to open properly so the urine can't drain away?

Medication for weak bladder muscles

There are medicines that can make the bladder muscle fit again and encourage it to get moving. The main active ingredient here is what's known as bethanechol chloride, which attaches to the receptors in the bladder wall and stimulates them. There are also other kinds of medications that are supposed to make your bladder muscle more toned. You should speak to your doctor and test out which active ingredient works best for you.

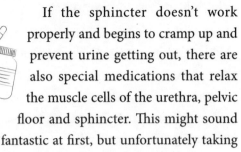

If the sphincter doesn't work properly and begins to cramp up and prevent urine getting out, there are also special medications that relax the muscle cells of the urethra, pelvic floor and sphincter. This might sound fantastic at first, but unfortunately taking them has a whole host of side effects: not infrequently, the entire muscular system is weakened, which can lead to overall tiredness and inertia. Gastrointestinal problems and a general feeling of sickness are not uncommon, either. It can also take a really long time until you and your doctor find the right dosage. Patience and resilience are required.

Teaching the bladder when to empty itself again

This works best through toilet training. In this process, those affected go to the toilet and pee at particular, predetermined times. And yes, they also do this when they don't need to pee at all. This is supposed to strengthen the bladder and get it accustomed to filling up normally again and then contracting when the time is ripe.

Bladder training might go like this, for example: you drink a large glass of water roughly every two to three hours, and you go to the toilet half an hour later — even if you don't feel the urge. This way, the bladder is supposed to learn how to behave normally again. Sure, it won't happen overnight. But don't worry, the bladder is a really good student. Just give it and yourself enough time to get a new and improved peeing rhythm.

Relax with pelvic floor exercises

There's also something more active you can do for yourself: pelvic floor training. But how are pelvic floor exercises supposed to help when nothing will come out of your bladder anyway? What's happening surely has nothing to do with a weak pelvic floor, right? True! But while strengthening the pelvic floor, you will also learn how to relax it. Reverse training, so to speak.

First up, it's pretty important that you have an idea of precisely where your pelvic floor is and how you can activate it. When you're peeing, just try interrupting the stream of urine. The muscle that you use to do this is exactly the one that you should be actively tensing and relaxing. (A word of warning though: don't do this too often! Otherwise you'll totally confuse your bladder and make the symptoms worse.)

Like with progressive muscle relaxation (pages 98–100), first you have to build up pressure in the pelvic floor and tense it. For this, do what you did on the toilet to stop peeing, but this time on dry land, as it were. Hold this state for about ten seconds and then actively relax the pelvic floor. Believe me, it's not that easy. To achieve results, you should do this exercise for twenty minutes a day. The good thing is that you can do this workout any time and any place. On the train, in the office, at the cinema, lining up at the supermarket, while writing a book …

Breathing properly while doing pelvic floor exercises

To train your pelvic floor properly, it's important to be aware of your breathing. The diaphragm — the most important muscle for breathing — bulges out when you inhale, pushing the abdominal organs and pelvic floor down. When you exhale, the opposite happens: the diaphragm lifts and the pelvic floor muscles contract and are pulled up. If we work against our breathing, that is, pull the pelvic floor in when we inhale, it's harder for the muscles to first contract powerfully and then to relax. So it should go like this: breathe out — tense the pelvic floor; breathe in — relax the pelvic floor.

If you've just been caught out exercising your pelvic floor exactly the opposite way, *i.e.* tensing it when you breathe in, now is the time for a change. If you find it hard, draw on your imagination again. Visualise the space in your stomach increasing with every exhalation. The diaphragm is making more room for your pelvic floor so that it can spread out nice and wide when it tenses.

Perhaps it will help you to visualise it. Your pelvic floor is a cloth stretched on a string. The string runs through your entire body, through your stomach, across your chest and ends in your head. If you tense your pelvic floor, it's as if this string is being pulled and the pelvic floor lifts up. Relax it again and the exact opposite happens: the pelvic floor slowly lowers down and unwinds. If you want to be more romantic about it, you can also picture your pelvic floor as a delicate flower opening and closing. Let your imagination know no bounds.

Along with active relaxation of the pelvic floor, warm baths are also good. Lie down in the tub, close your eyes and try to relax yourself and your pelvic floor. If you don't have a bathtub, you can also use a hot water bottle: lay it on your lower belly and perineum. It's amazingly comforting.

Draining the bladder yourself: an introduction to introducing catheters

If none of the above-mentioned treatment methods have worked and you still just can't go, it's time for you to haul the urine out of your bladder yourself, using self-catheterisation. To do this, you'll introduce a catheter into your bladder several times a day, which will drain the urine away for you. Yes, I know it sounds terrible at first.

The catheter is made of a thin, flexible silicone hose, which is usually covered with a lubricating layer intended to avoid irritation and so on — unfortunately, it's relatively easy to invite inflammation or an infection into your bladder with a catheter. To stop this from happening, it's important to have a clean technique. At what angle do you introduce the catheter tip so that you don't inadvertently poke the urethra? How far must the tip be inserted so that it sits properly? A specialist will explain all this to you in detail, so there's no need to be scared. Trust yourself, and ask a tenth or fifteenth time if you're unsure.

First you disinfect your hands and your genital area. For this, there are some tools you should always have with you from now on: tissues, disinfectant, lubricant. To disinfect your genital area, stand with your legs wide apart, carefully pull your labia apart and swab the upper part with a disinfectant wipe. Depending on whether you want to use lube or not, squirt some of it into the urethral opening. This way, it'll be easier to insert the catheter.

To do this, you now spread the labia again and pull them forward a little. This way you can see the urethral opening better, which is where the catheter is about to go. Now take the catheter out of its package and introduce the tip slowly and carefully into the urethra. If you're doing everything right, the urine should now start to flow. It will either land in a pouch attached to the catheter or into a separate container that you can empty into the loo.

Once you've finished tinkling, carefully remove the catheter from your urethra and dispose of it. The first few times you do it, your urethra might feel a bit raw or sore, but this will go away after a while.

If you get a UTI after beginning to do this, it's best to have it checked out by your urologist. And another tip: try to be as laid-back as possible about the whole thing. The catheter is now part of your daily toilet routine, so you should make friends with it. It might help to tell your family and colleagues about it, so you don't feel silly when you disappear into the toilet with all your equipment and are detained there for a little longer than usual.

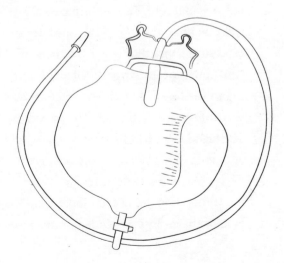

Other Reasons Your Bladder Might Act Up

I repeat: it's not always your bladder or urinary tract's fault when water storage and disposal doesn't go as planned. Often, the problem lies a few levels higher or deeper, that is, in the nerves or the vulva.

Neurogenic bladder

Remember, our nerves are supposed to forward the tinkling messages from our bladder wall to the brain. Sadly, though, they don't always do this conscientiously. This could be because of damage to the spine, or nerve disease. Specialists then call it neurogenic bladder.

You can imagine the whole thing like a three-storey factory: on the ground floor is the bladder. This is where you'll find the stretch sensors that sit on the bladder wall and track how big the bladder is getting as it fills up. They send these measurements on to the department on level two: the nerves in the spinal cord. They in turn pass the messages they receive right on up to the top, to the third department: our boss, the cerebrum. Once they've arrived there, our brain evaluates the situation: if we are currently in an important meeting or waiting in the line at the supermarket checkout, the brain sends this information through the nerves back to the bladder. The bladder muscle stays relaxed, the sphincter stays closed, and nothing is voided. As the couriers of these messages, the nerves play an essential role.

If they stop doing their job, *i.e.* they get stuck in a traffic jam, or they transfer encoded or incorrect messages — or no messages at all — the whole production line goes out of sync and the bladder acts up. While in our meeting or the line at the supermarket, we suddenly need to go to the toilet super urgently, and might even leak a little. Or the complete opposite happens: our bladder basically stops talking to us and we have problems getting urine to come out.

Which of these two bladder problems you suffer from often depends on where exactly the nerve disorder is. If the nerve damage is farther up — say, for example, if you have a prolapsed disc in the neck area — OAB or even urge incontinence (see below) might appear on the scene, or it might make you need to go to the toilet every twenty minutes at night. If, on the other hand, the damage is farther down, in the area around the lumbar vertebrae just above the bum, this is more likely to indicate a voiding dysfunction, so you might have problems emptying your bladder properly.

Surprisingly, people often don't even notice that they've injured their spine. Sure, for a day or two your neck might tweak or twinge a little, but you quickly forget about it. That's why it's very important that your urologist looks at your entire clinical picture really closely. If there is the slightest suspicion that your bladder problem might be neurological then the nervous system and the brain must be carefully examined along with the bladder, which may require an MRI scan.

A difficult bladder due to endometriosis

Endometriosis is the It Girl of gynaecological disorders. In recent years, no other condition has so often found its way into Instagram feeds, talk shows, news articles, books and

— yes — unfortunately even into the abdomens of women and girls. All of a sudden, lots of sufferers are coming forward and explaining endometriosis and its associated problems. No wonder: according to medical experts, endometriosis is the second most common gynaecological condition (the first is uterine fibroids, that is, benign growths in the uterus); world-wide, roughly 190 million women — that's about 10 per cent of women of reproductive age — have endometriosis.[12]

But what exactly is endometriosis? Let's take a little excursion to the womb. Endometriosis is when little clusters of uterine lining — aka endometrium — abandon the uterine cavity, go wandering, and make themselves at home on other organs for which they were not intended. Like so much to do with women's bodies, why they do this has not yet been sufficiently researched. Who knows, perhaps the ovaries or the rectum are just more exciting than the womb. Experts call these uterine escapees endometrial implants. These implants, or lesions, are usually influenced by menstruation and its associated hormones, which is when they can really grow and develop. The result? Severe period pain.

My tip: if you have been battling painful periods for a long time (and not just 'Oh well, time for a hot water bottle and painkillers' pain; I mean the really bad stuff!) then go to a doctor and have it checked out; there are even specialist doctors just for endometriosis. The pain can be so bad that women and girls who have it faint because their circulation is affected. Sex, along with gynaecological (and urological) examinations, can also hurt, or you might have pain in your lower abdomen and genital area even without touch or penetration. You should perhaps also know that endo — as its sufferers have baptised it — is one of the most frequent causes of infertility.

Going to the toilet can also be problematic, for both number ones and number twos. It entirely depends on which organs the endometrial legions — I mean, lesions — have camped out in. Some growths choose the bladder as their new home. According to experts, between one and five per cent of all cases of endo involve urinary tract endometriosis.[13] Here, the endometrial implants' favourite spots are the urinary bladder followed by the ureter, and, very occasionally, the kidneys or the urethra.

Those affected might realise that endometriosis has formed in their urinary tracts through recurrent bladder infections, pain in the flanks (that is, the lower back area) or even dysfunctional voiding of the bladder. These complaints are usually related to menstruation, *i.e.* they are worse during your period. To figure out whether endometriosis is lurking behind your bladder problem and you might have a bunch of stowaways in your urinary tract, you should get checked out by your doctor.

Problems with the bladder because of problems with the vulva

Ever heard of vulvodynia or vestibulodynia? No? Don't worry, you're not alone. It's hard to imagine what's behind these complicated terms, but they could be relevant for all you ladies who often suffer from bladder infections or are tormented by an irritable bladder. Roughly speaking, these terms are about pain in and on the vagina. Vulvodynia refers to pain in the entire vulva, that is, the vagina, labia and co., while with vestibulodynia, the pain is 'only' between the labia, that is, in the internal, moist part of the vulva. Sufferers have difficulty enjoying their sex life because any penetration hurts like hell. Even inserting a tampon turns into a painful activity.

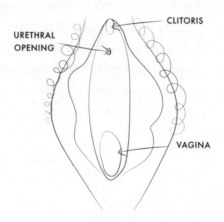

So what's all this got to do with the bladder? Along with the symptoms already mentioned, affected women often describe ones that resemble the typical signs of a bladder infection. In other words, a burning sensation when urinating, pain in the lower abdomen, a frequent need to pee, or blood in the urine.

So obviously your first bet would be a bog-standard UTI. Unfortunately though, when doctors don't find any bacteria that could be responsible for the inflammation through a urine test, these women are often diagnosed with interstitial cystitis (pages 79–80) and treated for that. But this won't fix the problem.

It may take many years and even more visits to the doctor until a patient is diagnosed with vulvodynia or vestibulodynia. This is also because lots of doctors are quite simply not (yet) well enough schooled in this area. Patients are often told, 'Unfortunately, we can't find anything; there must be a childhood trauma that you urgently need to address …' and the problem is chalked up to mental health. It's true that it could be psychological, but it might not be. Diagnosis of both vulva conditions happens through a classic process of elimination, and various tests will be necessary.

Women with these conditions often spend years trudging from practice to practice, where they are just fobbed off with antibiotics again (if they're given anything at all), or not really taken seriously. So explore all options available to you and don't let yourself be discouraged. If you often have the symptoms of a UTI but no guilty bacteria are detected in your urine, you should have a chat to your doctor about the possibilities of vulvodynia and vestibulodynia. Even if you don't find sex or inserting a tampon painful. Unfortunately, the conditions can't yet be cured, but there are several treatment options that at least offer relief.

The First Time:
Bracing Yourself to Visit a Doctor

Congratulations! You've sought professional help, and with this you've taken the most important step for your bladder health. So here's what will happen next.

Usually, the receptionist will immediately present you with a container for an initial urine test, which will look for possible bacteria, viruses or other culprits, and more generally for substances that really have no business being in your pee. Then comes the first consultation.

Here you will be able to tell straight away whether you've got a good doctor or not. How is the communication? Do they ask you detailed questions and write notes? And, the number one most important thing: do they take you seriously? If not and your doctor sneers at your urinary issue as a 'typical woman's problem', tells you that 'it's just in your head', or wants to palm you off with a variety of pills, you should definitely find a different doctor. I'm sorry to say that I'm speaking from experience here.

First and foremost, it's important that you trust your doctor. If you don't feel comfortable, don't shy away from visiting another practice, and, before this, get recommendations. If you are a rookie in the world of bladder problems and don't (yet) have a urologist on your team, you might want to talk to another trusted doctor first.

At the start of the examination, you will be asked about any possible pre-existing conditions. Do you often struggle with back pain? When did you last have surgery? This will give your doctor an initial picture of you and your body and help her rule out or take a closer look at certain things. Then she will take a detailed medical history of your pee problem.

Get ready for the following questions:

'When did the symptoms first appear?'
'Did anything happen at the time?'
'How often have you involuntarily lost urine?'
'Roughly how much leaked?'
'Do you wear incontinence pads for protection?'
'If yes, how often do you have to change them?'
'Do you have problems with your bowel movements?'
'What is your sex life like?'
And so on …

Of course, these questions might feel uncomfortable and will take some getting used to, but try to answer them thoroughly.

After the medical history will come an examination of your kidneys, bladder and reproductive organs. By touching and pressing on them, the doctor can check if the organs are in the right position, if they are swollen, or if the skin around them is reddened. The bladder, by the way, can be felt from the outside at a volume of 150 millilitres, and at 500 ml it

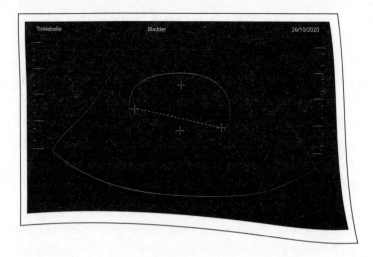

bulges out so it is visible to the naked eyes (depending on how much belly fat is in the way).

To get more detail, the organs are then assessed again via ultrasound. How big is your bladder? How thick is the bladder wall? Can any urinary stones be detected? Or cysts, perhaps? Are the kidneys working properly? Are both of them functioning equally well, can the urine flow normally and so on. Since your bladder will usually be empty for the examination (after all, you did a urine test beforehand), the urologist can now determine via ultrasound if it was actually voided properly, or whether there is residual urine.

For the examination of your reproductive organs, you'll lie back or take a seat with your legs spread. The urologist will first attend to your urethra along with the urethral sphincter muscle and pelvic floor, and they'll do this either with their fingers or with what's called a perineometer. This is a small device a bit like a dildo that's inserted into the vagina to measure your pelvic floor muscles. Then you should be checked for a possible prolapse of the uterus, bladder or vagina.

And, finally, we get down to the nitty-gritty tests, such as a pressure measurement, or a swab from the urethra. This might make your heart or your bladder a little uneasy. Don't be scared, though. We'll go through the most common tests step by step.

The urethral smear

The urethral smear is similar to undergoing cervical screening — it's just that it's not the cervix and uterine orifice but the urethra that is examined. Your doctor might be looking for bacteria such as ureaplasma or chlamydia, which can indeed also cause problems in the bladder. To do this, the doctor will have to go 'deeper' with the cotton bud: thorough swabbing or scraping is needed to ensure they end up with enough tissue to examine. Yes, it's uncomfortable, and yes, it hurts. Unfortunately, you just have to say to yourself 'Close your eyes, open your urethral sphincter and get on with it,' and remember that it will help you in the long run.

You and your bladder will be quite worn out after the swab, so there might be some mild pain afterwards when you pass water, and don't be alarmed if there is even a bit of blood.

Urodynamics

If this smear reveals nothing that might explain your bladder problem, it could be appropriate for your doctor to do what's known as urodynamic testing. This might happen in the practice itself, or you might be sent to a clinic that has the necessary equipment. In urodynamics, doctors look more closely at the lower urinary tract. Roughly speaking, the test measures bladder pressure, *i.e.*, when and how strongly the bladder muscle contracts. Before it starts, a urine test is used to check whether you could be suffering from an infection — if so, the examination can't proceed.

First of all, two thin catheters — or probes — will be inserted into you. One in front through the urethra into the bladder, and one into the rectum from behind — yep, through your backside. These measure the pressure that is put on the bladder from the outside when, for example, you squeeze or cough. Sensors will also be stuck to your skin in order to track the muscle contractions of the bladder and the internal urethral sphincter. Next, your bladder will be filled slowly with lukewarm saline solution. At first you won't notice it, but the fuller the bladder gets, the stronger the urge to pee will become. The sensors on your skin can now measure how much your bladder muscle contracts as the volume of saline solution increases, thus increasing the pressure on the sphincter.

This sort of examination usually happens on a toilet-seat-like device, so the 'urine' (the saline solution) you can't hold in runs into a bucket instead of onto the floor or a seat. This test can also be used to check whether you lose urine unintentionally and, if so, how much. Measurements are recorded throughout the entire examination and then evaluated. In the end, your doctors will know exactly how your bladder behaves as soon as it starts filling up.

Uroflowmetry

Uroflowmetry measures the force and speed with which your urine leaves the bladder. This test is done to detect any possible urethral narrowing. Between you and me: this test is the easiest. All you have to do is pee. You do this into a special toilet that measures the force of your urine stream. After this, the doctors know exactly whether you're tinkling normally, or if you're passing water too quickly or too slowly.

Cystourethrography: X-raying the urinary tract

If you get shy peeing in the presence of strangers, try doing it while you're being X-rayed. To ensure that the doctors can really see everything, a catheter is used to inject X-ray contrast dye into your bladder. This way, what happens to the urine in your bladder can be closely monitored. Does it flow back into the ureters? Is residual urine left behind? All of this can be checked precisely with voiding cystourethrography.

Cystoscopy

If your doctors aren't sure what's affecting your bladder, it makes sense for them to take a look at it from the inside, by performing a cystoscopy. Here, a thin, flexible tube with a camera at one end, known as a cystoscope, is inserted into the urethra and used to take a look around your bladder. Cystoscopies are usually done under local anaesthesia. This means that the entrance to your bladder is made numb with an anaesthetic gel. And girls, this is where our short urethra is a plus: for us, a cystoscopy is quite a quick, easy and painless affair.

As with a urethral smear, though, your urinary equipment will be quite irritated and battered after this examination, so burning and sensitivity are normal. If this strange and uncomfortable feeling lasts for longer than a few days, please inform the practice immediately.

After any of these tests have been done, your doctor will explain the results to you and you can discuss what's next for you and your bladder. Which medications will help? Will pelvic floor exercises be beneficial? Or are further tests necessary? If your urologist suspects your bladder problem has been caused by damaged nerve pathways, you will be referred to a neurologist to have your spine or head checked using an MRI or CT scan. To test whether your kidneys are still working properly or whether they might be impaired, a radiologist can perform what's known as renal scintigraphy. Here it's not your spine or head that is screened but your kidney function: are the kidneys receiving enough blood, are they doing their job properly and transferring the urine on to the bladder, and so on. You will receive further treatment based on the results of the tests.

The bladder diary

Before you embark on any of these procedures, the first thing your urologist is likely to want from you is what's known as a bladder diary. But before you start writing 'Dear Diary, my bladder drove me crazy again today,' what we're dealing with here is actually just an Excel table that you're supposed to fill in with facts and figures; it's also called a frequency volume chart.

Through the bladder diary, which I've lovingly christened my pee papers, you can determine abnormalities in your drinking and toileting behaviour. To make sure it works properly, you list exactly how much you drink, when you go to the toilet, how much urine you void, how extreme the pressure was, whether perhaps anything leaked out, and so on. To do this you are going to need a new toilet buddy: a little cup with which to measure the exact volume in millilitres of your urine.

Yep, it's pretty annoying and arduous. And, of course, the whole thing is even harder if you work in an open-plan office with shared toilets. When you vanish off to the toilet every twenty minutes with a little measuring cup and then frantically enter something into a notebook or spreadsheet, it's only natural that your colleagues will eventually start to wonder. Make it easier on yourself and tell your favourite workmates about your pee problem. Uncomfortable questions will automatically melt away and you'll no longer feel like you're being watched.

How long you'll need to keep this bladder diary depends entirely on your doctor. At the start, about three days should be enough to reflect on how often and how urgently you need to go to the toilet. With the help of this information, your clinical picture can now more easily be assessed. It can be very informative for you to see in black and white how much you drink and how often this makes you need to go to the toilet.

Another benefit is that you might also be able to draw some conclusions about which drinks send you to the toilet more often. If you then start taking medicine that is supposed to help limit your need to pee, you can use the chart to better recognise how well it's working.

Tinklebelle's Bladder Diary

Day 1 Date ..../..../....   Day of the week.............

TIME	URINE volume in ml	FLUID INTAKE volume in ml	what	UNVOLUNTARY URINE LOSS

Alternative Treatment Options

Have you put your bladder through its paces, enduring practically every test and treatment Western medicine has to offer, without anything really helping? Perhaps you should consider other options.

Psychological help

Understanding the language of the body isn't always easy, even though we live in it. Sometimes, the body may try to tell us something that our mind hasn't acknowledged, for example, that we are stressed, anxious or sad.

How exactly it does this differs from person to person. Some people suddenly get headaches, insomnia, or indigestion. Other people transform back into their pubescent selves and have to deal with irritated skin and really nasty acne. And other people's bladders go crazy.

As my osteopath once said: 'Your body does everything to protect you.' And that can include expressing mental stress that we haven't dealt with yet. If you think this could be you, it's definitely worth talking to your doctor and getting some psychological help if necessary.

Along with this, it also makes sense to take life at a slower pace. Inhale deeply and exhale, meditate, do yoga, try painting or pottery, or simply go for a walk and listen to the birds — whatever you find relaxing. Try to listen to your body and its needs.

Traditional Chinese medicine

According to traditional Chinese medicine, or TCM for short, problems with and in the bladder often arise because the Yin and Yang — our female and male energy — aren't in harmony. TCM suggests that disharmony can also develop in the bladder or kidney Qi, *i.e.* in the specific energy of these organs. To treat this, traditional Chinese medicine uses specific medicinal herbs and/or acupuncture. Don't be scared; despite the use of tiny needles, acupuncture is a largely painless treatment.

Acupressure

Another approach from TCM is acupressure. Unlike in acupuncture, here pressure is applied to specific points on the body using just the fingers. Like in acupuncture, this is supposed to allow energy to flow in a manner that will alleviate or even heal physical problems.

Acupressure and acupuncture points are located in completely different parts of the body from where the related physical problem may be occurring. The bladder meridian is the collective name for the bladder's acupoints, and it covers quite a large area: it starts next to the inner corner of the eye and extends over the head to the neck, and down to the back. From here, it continues to run in two separate strands along the back of the leg to the feet, and finally ends on the outside edge of the foot, on the nail of the little toe.

An acupressurist can of course treat your entire meridian for you, but you can also take care of it yourself at home for free. The point 'urinary bladder 60', for example, which is found on the outside of the ankle between the lateral malleolus (the knobbly bit on the outside of your ankle) and the Achilles tendon, is very important for bladder health. A little bonus: applying pressure to this point can also relieve head and neck pain. Simply massage this spot for about ten seconds by using your index finger or thumb to slowly and evenly increase pressure on it. Then release it, wait for a while and then do another pressure massage on the same spot. You can do this for as long as you have time and inclination, but at least five times should be enough. It's best to consult your acupressurist for more advice.

Osteopathy for bladder problems

Osteopathy is an alternative medicine. Its practitioners search for the source of pain in the body and then treat their patients using their hands. According to osteopaths, the bladder — like every other organ — needs enough space and room to move if it is to feel good and be able to work at its full capacity. If the bladder is crowded and restricted by other organs, adhesions, fascias, tendons or muscles, it feels pressured and goes on strike, and we in turn suffer from voiding dysfunctions, overactive bladder, or we might get UTIs more often.

Osteopaths try to straighten things out for the bladder using palpation with the hands. To investigate the cause of the bladder problem, osteopaths examine the anatomical and physical condition of each body part and organ. The culprit could be anywhere: in the jawbone, for example. If the jawbone is out of position, the pressure can have an effect on the pelvis, which in turn increases pressure on the bladder. Or your knee, which is linked to the pelvis via connective tissue in the upper leg, and can thus cause misalignment, which over time stresses the bladder and can trigger functional disorders.

5.
Home and Dry?
Bladder Weakness
and Incontinence

Incontinence … wow, what a taboo topic! Talking openly about the fact that you might not always manage to get to the toilet in time? Out of the question. The last time you 'officially' peed your pants was when you were in kindergarten or primary school. Back then, a grown-up usually comforted you tenderly — no need to be ashamed. And now? These days, you're mortified that every now and again some urine leaks into your pants instead of into the bowl. But you are not alone!

It is estimated that roughly 303 million women worldwide suffer from urinary incontinence.[14] And it's not just grandmas and grandpas who are visiting drugstores and surreptitiously sneaking value packs of nappies through the checkout. A 1995 review of several studies found that, in the general population, 20–30 per cent of young adult women, 30–40 per cent of middle-aged women, and 30–50 per cent of elderly women suffered from incontinence.[15]

It's hard to say how many people out there are really

affected by incontinence and bladder weakness, though, since it is presumed that the number of unreported cases is quite high. Why? Well, because it's thought that not even half of the leakers out there address the problem by visiting a doctor to get it treated. The shame of having to 'out' yourself is simply too great. According to a study by the medical and personal hygiene manufacturer Paul Hartmann AG, 39 per cent of sufferers have never talked about the problem with their partner, and as many as 45 per cent of German respondents avoid sex because of it.[16]

It's time to finally break the silence. Because yes, a leaky bladder can be treated. You just have to know how. And in the meantime, to stop you being afraid of a public mishap, it can be really helpful to rely on absorbent pads.

There are various kinds of incontinence, and while they are really quite different, they all have the same end result: wet pants. The most common forms are stress incontinence, urge incontinence and a mixture of the two, which is known as mixed incontinence.

Stress Incontinence, aka the 'Oops' Moment

With stress incontinence, the problem is not the bladder itself, but the pelvic floor. The external urethral sphincter and surrounding musculature of the pelvic floor are too weak and can no longer withstand even mild pressure from the bladder. This means that those affected lose urine without having sensed the need to pee — a few drops or entire loads of urine simply land in their pants without prior warning. This happens especially when doing physical activities like lifting or jumping, but a little bit of urine can escape even when doing the smallest things like sneezing, coughing or laughing. If the incontinence is really pronounced, sufferers even lose urine when they're lying down, *i.e.* when they're in a completely relaxed position.

Nature demands more of women's pelvic floors than men's, and this is why we are affected by stress incontinence more often. Unfair, right? It's because the female pelvic floor holds and supports important organs like the uterus. During pregnancy, the pelvic floor becomes softer and more elastic, and birth exacts a high price on it. When levels of oestrogen later decrease during menopause, the pelvic floor is no longer supplied with as much blood, making it weaker.

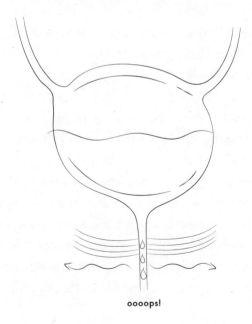

oooops!

Age gradually takes its toll on our pelvic floor. Between our 30th and 80th birthdays, the muscle fibres in our urethral sphincters shrink a full 65 per cent. In other words, we have less than half the strength we once had downstairs. Wrinkles on your face or a not-so-pert butt are tiny inconveniences in comparison. Apart from this, weak connective tissue and carrying excess weight can also play a big part in stress incontinence.

If the pelvic floor is too weak, there's another important task it can no longer really manage: supporting the organs above it. The technical name for when this happens is an organ prolapse, and it can also lead to bladder problems and incontinence.

Bladder problems and incontinence because of a prolapsed uterus, bladder, or other organ

The organs that really like slipping down are the uterus, the rectum and the bladder itself. And why do they do this? Our organs are normally held nicely in place by retainers such as ligaments, connective tissue and the pelvic floor. But experiencing pregnancy and childbirth, having generally bad connective tissue, being overweight or going through hormonal fluctuations can make these restraints looser and weaker so they're not as good at supporting the organs. The result? Danger: high risk of falling! Symptoms of this descent can be lower abdominal and back pain along with a feeling of having some sort of foreign object in you that is gently pulling down. Kind of like a full tampon that really needs to be changed.

Because the organs are knitted together by the restraints,

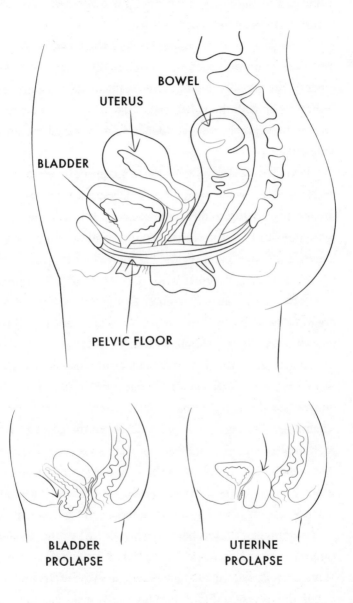

it means that once one organ slips, the others will soon slip too if the problem isn't treated.

If the descent is not remedied and the pelvic floor gets more and more worn out, one of the affected organs can fall down so far that it bulges out of the vagina and you can see it from outside with the naked eye. Doctors call this a bladder, uterine or vaginal prolapse (depending on which organ is affected).

As you've guessed by now, these prolapses have a considerable impact on our bladder and pelvic floor. The bladder prolapsing can lead to difficulties passing or holding water. If it happens to the organs that usually sit higher up, like the uterus, they then press on the bladder when they descend, resulting in dysfunctional voiding or irritation. According to experts, every second woman that suffers a mild pelvic organ prolapse has to contend with a weak bladder and the incontinence that goes along with it.

To determine whether your bladder problem is caused by a prolapse, your doctor will feel your pelvic organs from the outside and through the vagina. This way, she can determine whether everything is in its right place. So that she can also examine the organs at the back of the pelvis properly, your doctor will also do a short palpation of the rectum by poking about in your backside with her finger. (Yes, in a urology practice, it's not just men who get a thorough 'ass-essment'.)

If your pelvic floor is being pushed down by other organs, the problem can be set straight with what's known as a pessary. This is sort of like a silicon tampon in the shape of a cube, ring or cup that is inserted into the vagina. Because of the special shape and fit of the pessary, a supportive vacuum is created in the vaginal wall, preventing it from sinking. In terms of how it feels, a pessary like this is just as comfortable as a tampon or a period cup — you'll barely notice it, if at all. You can wear the pessary for as long as you want, but you should definitely remove it and clean it thoroughly when you go to bed.

Apropos of removal: this works a little differently from a tampon or a period cup. Because of the vacuum that's been created, it's more fixed in place and can't just be yanked out. (Warning: acute danger of pain and injury!) Before you take the pessary out, you should insert a finger into the vacuum and carefully move around it. This releases the vacuum, and you can now more easily remove the pessary. More severe prolapses can be treated with various operations.

Which of these different treatment options will help you the most depends on how old you are, whether you have had children and how bad or advanced your prolapse is. Your doctor will be the best person to advise you.

To stop it from getting this far in the first place and to keep your organs sitting where they belong, pelvic floor exercises really are invaluable. The better trained and cared for this mighty sling carrying all your organs is, the more secure your uterus, your bladder and their friends will be in your body.

Pelvic floor exercises

The good thing about stress incontinence? You can get a really good handle on it through targeted pelvic floor exercises. Building up the pelvic floor muscles increases their strength and thickness and improves co-operation between the nerves and the muscles. They can help you to keep the urethral sphincter more tightly closed and withstand bladder pressure again, without leaking when you cough, sneeze or jump. Trained physiotherapists can explain the best exercises for bringing your pelvic floor to heel — or rather, to heal. There are even specialist pelvic floor physiotherapists, and it's worth seeking them out.

More control with biofeedback

Biofeedback is a therapeutic method intended to help patients learn to better perceive physical processes they are usually not conscious of. This lends itself really well to the pelvic floor. You work with a little probe, which you insert into yourself so it can show you through vibrations, beeps or blinking lights exactly how much you should tense and then relax your urethral sphincter. If you do the biofeedback in a clinic, you can observe your pelvic floor tensing and relaxing live: the probe is often connected to a screen, and a kind of sine wave shows how strong or weak the tension is. It's kind of like a video game, which makes the otherwise quite dull pelvic floor exercises a bit more fun and exciting.

Hold it: vaginal cones

There are also vaginal cones that you can insert into your vagina and carry around with you. To stop these cones from falling out, you instinctively tense your pelvic floor and voilà, you're exercising it. The tampon-like cones come in different weights ranging from 20 to 100 grams, so you can keep training your pelvic floor to be stronger and stronger. It's just like going to the gym: you don't start with the heaviest dumb-bell; instead, you work your way up from lighter weights until you're practically a champion bodybuilder.

Energise with electrical stimulation

Pelvic floor training is really good when combined with electrical stimulation. The cool thing about it? You don't have to actively do anything at all. Just insert the electric probe and turn it on, then lean back and relax. The electrical impulses activate the various muscles of the pelvic floor, making them contract. For optimum training against incontinence, the pelvic floor should be stimulated with frequencies of between 20 and 50 Hz for a duration of 0.1 to 0.3 milliseconds. The probe will vibrate and tug a little, but apart from that you won't feel anything. For the best results, work your pelvic floor with the electrical stimulator one to two times a day for 15 to 20 minutes at a time — a perfect accompaniment to your favourite TV show.

Medication for stress incontinence

As a supplement to pelvic floor exercises, there are also medications that can ameliorate stress incontinence. If your pelvic floor is a little rusty due to potential hormonal fluctuations and a lack of oestrogen, hormone pills inserted vaginally can help. This should rebuild and spur on the mucous membrane.

There are also medications with duloxetine as their active ingredient, which are used for depression as well as stress incontinence. Duloxetine works centrally on the spinal cord, where it supports the pelvic floor via the nerves by increasing the concentrations of serotonin and noradrenaline.

Last resort: surgery

If these therapeutic approaches aren't working and pelvic floor exercises, electrophysiotherapy and medication haven't done anything, stress incontinence can also be treated surgically. Most often, slings are used: they are placed under the urethra, and function like a kind of hammock making the urethra really nice and comfortable. Stabilising the urethra takes pressure off the pelvic floor, which ameliorates incontinence. Your doctor can talk to you about whether this kind of operation might be suitable for you.

And by the way: if something goes skew-whiff when you physically exert yourself, it's no reason to worry. When the body is working flat out, it's quite common for continent women to leak a little, too. Survey results suggest that over half of elite athletes and dancers have experienced leakage in certain situations, and 60 per cent reported occasionally wearing panty liners to protect themselves from these little mishaps.[17]

Another study found that a full 80 per cent of trampolinists reported leakage during training.[18]

'It's Urgent!' Urge Incontinence

With urge incontinence, you lose urine involuntarily because the need to go is just too great. This is despite the fact that the bladder is not even full yet. Unlike stress incontinence, urge incontinence has nothing to do with a weak bladder that can't hold in urine. Here, too much power emanates from the bladder. The bladder muscle suddenly contracts with such force that even the strongest sphincter couldn't withstand the pressure, so it gives up and lets the urine flow. Imagine it as an attack on a fortress. Inside, they try to withstand the onslaught for as long as possible, but outside, bigger and bigger cannons

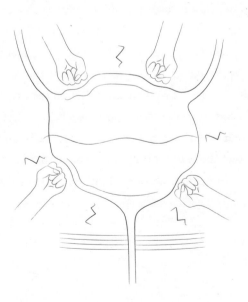

are being fired. Then the battering ram is brought out. The gate — in our case the pelvic floor — eventually gives out. A horde of invaders storms the tower, and we wet ourselves.

Annoyingly, urge incontinence often steps onto the battlefield with reinforcements: she brings along her greatest ally, the irritable bladder. This means that sufferers have to be careful to avoid wetting themselves not just every two or three hours but sometimes as often as every twenty minutes. And, of course, it happens at the most inconvenient moments: at the supermarket checkout, during a speech, or on a date at the park. All rather tricky situations for a turbo sprint to the nearest toilet.

There are various reasons why the bladder muscle might suddenly start convulsing.

The bladder receptors are to blame

The receptors that measure bladder pressure can act up. It's like they start lying to the nerves and broadcasting fake news about the level of liquid in the bladder. The brain receives this misinformation and gives the order for immediate evacuation. The result: we have to go to the toilet super urgently and eventually wet ourselves if we don't reach the toilet in time. As we have learned, this can be caused by a lack of oestrogen, which throws the bladder off balance; or an unhealed bladder infection or interstitial cystitis might have damaged the bladder wall.

The problem is with your nerves

As with all bladder problems, the culprit here could of course also be your nerves: they may not be transferring the signals from the bladder sensors to the brain properly. Our brain keeps receiving the signal that the bladder is full, even though it actually isn't at all, so the bladder muscle contracts and we need to steer for the next toilet ASAP or we'll have an accident. As we have seen, the reason for the nerves sending faulty information could be damage to the spine, such as a slipped disc, or nerve diseases such as Parkinson's or multiple sclerosis.

Treatment options vary depending on what has led to your urge incontinence. If the problem is with your nerves, the underlying illness must first be treated. Whether this happens through physiotherapy, massage or even surgery is something your doctor will discuss with you. If you have urge incontinence because of a UTI or interstitial cystitis, you should seek out the best treatments for these conditions. If your bladder's urge problem is down to an oestrogen deficiency, doctors can help with hormonal treatments, which are often administered as suppositories.

Medication for urge incontinence

If no explanation can be found for your pee problem, various medications are available that can relieve and relax you. As in the case of overactive bladder, anticholinergics can be used to prevent the receptors in the bladder making it contract, or Botox treatment or even a bladder pacemaker can also be considered.

Exercising the pelvic floor is always good!

Although urge incontinence has nothing to do with your pelvic floor or sphincter being too weak, training the pelvic floor is never a bad idea. The stronger your pelvic floor, the more power the bladder muscle needs in order to give you wet pants. Apart from this, a strong pelvic floor gives you a better sex life and ensures a better, more confident posture. So: work it, ladies!

When Two Become One:
Mixed Incontinence

Why settle for just one kind of incontinence when you could have two, right? Mixed incontinence is an annoying combination of urge and stress incontinence. In other words: sufferers are leaky down below because the pelvic floor is too weak, but they also experience superstrong, extreme, and sudden urges to urinate, because the bladder muscle is too strong and forces the urine out.

Usually, the more pronounced form of incontinence is treated first. That doesn't mean you have to resign yourself to the other; it's just that, sadly, parallel treatment doesn't tend to work.

So, if you more often wet yourself because your pelvic floor doesn't have a tight seal, *i.e.* you suffer from stress incontinence, the focus of the therapy will be there. Once this improves, the lesser (but still really annoying) urge incontinence will be dealt with. Unfortunately, though, some treatments can cannibalise each other. In other words, a treatment that relaxes the over-strong bladder muscle can do the same thing to the urethral sphincter, thereby having a negative impact on stress incontinence.

Constant Dripping: Overflow Incontinence

With overflow incontinence, the bladder becomes so full that it leaks and is practically always dripping. Unfortunately, owners of these overfilled bladders are incapable of emptying them normally: when they go to the toilet, all that comes out is a little trickle and residual urine is always left behind — technically, it's a voiding dysfunction (page 108–11). This is not only extremely uncomfortable: it can also be really dangerous. The residual urine offers the perfect breeding ground for bacteria and germs, which can in turn cause inflammation. One of the most frequent causes of overflow incontinence is an enlarged prostate, so — unlike stress incontinence — men are more commonly affected than women.

When the Bladder Is Innocent: Reflex Incontinence

Here, the co-ordination between the brain, the bladder muscle and the sensors on the bladder wall stops working properly, so the detrusor contracts uncontrollably. In contrast to urge incontinence, reflex incontinence — as the name perhaps suggests — always has a neurological cause. For once, it's not actually the bladder's fault.

Not Funny: Giggle Incontinence

Laughing is good for you; less so when it makes you wet your pants. But this is exactly what happens with so-called giggle incontinence: a sudden fit of laughter causes extreme bladder pressure and the normal micturition reflex is interrupted, so sufferers can no longer hold onto their urine. Or, as I like to say: I laughed so hard, tears ran down my legs …

Although this kind of incontinence is very rare, it mostly affects young girls during puberty. As if there weren't enough challenges to being a teenage girl! Since giggle incontinence is a kind of urge incontinence, it is treated similarly.

Coital Incontinence:
When You Feel Something Coming and It's Not Just an Orgasm

You're really getting into it in bed, the crucial moment arrives and you are just about to orgasm. But uh-oh! Something else is coming along with you. Before you know it, suspicious moisture is spreading all over the sheets — or your partner. Yes, you have just wet yourself during sex.

Firstly, try to calm down and don't panic. Roughly every fifth woman has lost urine during sex, either while climaxing or in the lead-up to it, during penetration. Unsurprisingly, not many people feel able to speak openly about what's known as coital incontinence: putting two very personal subjects together (sex and urination) is a recipe for taboo.

We're not talking here about female ejaculation, which is thought to happen when a secretion from the Skene's gland is expelled. More research is needed into how this abrupt expulsion of ejaculate, which can feel so much like urine, and actual coital incontinence relate to each other. How many female bodily phenomena have been taboo for centuries? But generally speaking, if you secrete more than a cup of fluid during sex, you're most likely dealing with coital incontinence.

Coital incontinence is mostly either down to the pelvic floor being too weak or the bladder muscle being too strong. That's right, our two favourite forms of incontinence — urge and stress incontinence — are once again responsible for our misery. During penetration, the pelvic floor is highly stimulated and supplied with lots of blood, so the bladder can

YEEEAAAA!

accidentally receive the impulse to pass water, and lets rip.

Another possibility is that a particular sex position is putting a lot of pressure on your bladder. If the pelvic floor is too weak, urine is then pressed out. That's why the first step is to try out other positions. Regular pelvic floor training should also be on your agenda. This will help your sex life and your ability to orgasm. If you don't notice an improvement, discuss the problem with a doctor.

Incontinence:
What to Expect at the Urologist's

I've got good news for you: there is help out there, if you're ready to go get it. You don't have to silently grapple with wet underwear, nappies or constant anxiety for your whole life. But you do have to leave embarrassment at the door of your doctor's office and be really open in explaining how your incontinence manifests itself. When did the problem first crop up? How pronounced is it? Do certain situations make it better or worse?

The examination afterwards will be exactly like the one for bladder dysfunction (page 128). Your urine will be screened for potential infections that might be causing your suffering. Then there will be a physical examination of your sexual organs to test how strong or weak your pelvic floor is, and whether there are any prolapses. Next, the entire urinary tract, *i.e.* the bladder and kidneys, will be scrutinised via ultrasound. Your lovely bladder diary will also be used here: you'll record exactly how much you've had to drink, when you've needed to go to the toilet, how much came out when you did, and — really importantly — whether you managed to get there in time. This way, both you and your urologists can see how severe your incontinence is. Your doctor can use a urethral swab to determine what shape your urethra is in and whether it might be affected by an oestrogen deficiency, which could be the reason for your incontinence.

How much urine do you really lose? The pad test

A pad test studies the liners designed to protect you against leaks, to see how much urine is lost over a particular period.

First up, a dry pad is weighed. Then you wear the pad while being put through a rigorous regime: you drink a lot, move a lot, cough, jump and so on. You can't change the pad while all this is happening. After two hours, the pad is then weighed again. The difference in weight provides precise information about how much urine you've lost in this time.

If you're worried that you might have stress incontinence, it can be measured using a stress test. Don't worry, it sounds worse than it actually is. In a stress test, you lie on your side on an examination table and cough on command. If urine lands on the table involuntarily, the test is positive and you are suffering from stress incontinence.

To investigate and assess the cause of the incontinence, your bladder will be looked at more closely through urodynamic testing. Doctors can use this to get a more accurate picture of the sensitivity, muscle strength and volume of your bladder. A catheter is used to fill the bladder with warm saline solution, and electrodes measure at what point and how urgently you start needing to pee. There's more on urodynamics on pages 131–2.

The Four Levels of Incontinence

Incontinence symptoms can be divided into four different degrees of severity to help in assessing the impact on the life of the patient, their treatment options, and how urgently they should be implemented.

Level one, the mildest form of incontinence, is when individual droplets or a little stream of urine leaks out. Usually, no more than 40 to 300 ml flows into your pants.

At level two, moderate incontinence, there is more leakage and it also happens at night. Between 300 and 1,000 ml of urine is lost per day, so highly absorbent pads can still provide good protection.

This is no longer true for level three, severe incontinence. You reach this level when the bladder completely empties itself without your say-so several times a day and at night. Here we're talking about a volume of 1,000 to 2,400 ml per day.

The worst form of incontinence is when the sufferer cannot control their bladder themselves at all anymore, and urine virtually just passes through it.

With incontinence levels three and four, those affected can either no longer leave home at all, or only wearing a nappy.

TIPS FOR INCONTINENCE SUFFERERS

• Don't be scared to address your problem openly and truthfully. You're not the only one going through this. It's really important that you find doctors you can trust.

• Always be prepared when you're out: pack leggings, socks and lightweight shoes such as ballerina flats. This way, if you don't make it to the toilet in time, you can quickly get changed.

• If you don't want to or can't take a change of clothes with you, it's better to reach for a long coat than a short jacket. This way you can hide your mishap on the way home.

• Familiarise yourself with the place you're going to in advance. Where are the nearest toilets? Are they easy to get to? Will you have to wait for long?

• Don't be ashamed of using absorbent pads or even nappies.

• Wet spots are less noticeable on black or dark trousers.

• If you know that you won't be able to go to the toilet for a while, adjust how much you have to drink.

• Try to avoid diuretic food and drinks such as coffee and asparagus altogether — switch to alternatives instead.

• Take every opportunity to pee that comes your way. This also helps on a purely psychological level, because you know that your bladder has just been emptied and can't start whingeing again in five minutes.

- Try tips such as kneeling or sitting down to escape your enormous urge to pee, and kill a bit of time until the next toilet (more on this on page 103).
- No matter what kind of incontinence you suffer from, pelvic floor exercises are always good. The more secure you are down below, the better.
- If you can't wait any longer, don't be ashamed to go behind a bush or a wall. Men do it all the time. There are products that might be able to help you with this, such as a Urinella, Shewee or similar. And, if people see you, think about this: what bothers you more? Someone giving you an irritated look for five seconds and then forgetting about you, or a ruined day/evening because of wet pants? Exactly.
- Don't be too hard on yourself, your bladder or your sphincter. You're doing the best you can. Body shaming is so last year.

When You Wake Up and the Bed Is Wet: Enuresis

If you wet the bed now and then as a kid, it's annoying (mainly for your parents, that is; after all, they're the ones who have to change the sheets), but it's not the end of the world. You usually grow out of it (more on this on pages 36–8). But imagine this situation as an adult. While other people are torn out of their sleep by a shrill alarm, for people affected by enuresis it's their wet mattress that does it.

Bedwetting is a huge psychological strain for those affected and their families, and it can have several causes. It is often quickly put down to a mental health issue, and while it's true that the problem could be psychosomatic, there are definitely also physical factors that can be responsible. For one, sufferers may produce too little of the urine-curbing hormone ADH, resulting in too much urine being produced at night. Remember: ADH is also slowed down by the consumption of alcohol, making us need to go to the toilet more often when under the influence.

It's possible that the interaction between the bladder muscle and the nerves of those affected is at fault. Another possibility is that you just sleep so deeply that you don't notice when your bladder is full and requires emptying. If this is the case, you should definitely swing by a sleep laboratory and get it checked out. Other things that could be lurking behind enuresis are our old friends urethral strictures, and bladder stones.

If enuresis is affecting you, don't be ashamed or hide away; let experts help you. They're usually very good at what they do.

The House Door Phenomenon, aka Latchkey Incontinence

Hands up if you've ever experienced this: you're on your way home, and your bladder is putting on the squeeze. Racing to the bathroom, you finally make it to your front door and are almost in sight of the toilet, rummaging around in your (much too large) bag for your keys, when the urge to pee suddenly gets so strong that you feel like you're about to have an accident. Sometimes you lose the race spectacularly, right there at the finish line.

Urologists recognise this problem as a kind of 'situationally triggered urgency incontinence', and it is often referred to as 'key in lock syndrome' or latchkey incontinence. Whatever you want to call it, it's uncomfortable and annoying; it can even make people anxious about being far from a toilet or lead to feelings of isolation and depression. So why does it happen? Why does the bladder suddenly run amok when it must know it's about to be emptied anyway?

On the one hand, the phenomenon can be explained by the pressure and stress that develops when you're out for a long time without being able to empty your bladder. It behaves like you're about to have a job interview: the nervous system activates the bladder at the worst possible moment (you'll learn more about this on page 189 below).

On the other hand, it could be force of habit, or the result of the Pavlovian reflex — that is, the classical conditioned response. Perhaps you've heard of Pavlov's dogs? Well, the Russian doctor and early behavioural scientist Ivan Pavlov noticed that his dogs didn't start drooling when they started eating, but when they saw the approaching white lab coats of the people who fed them. To corroborate this, Pavlov started sounding a bell every time his dogs were fed. And I'm sure you know the punchline: the dogs started to drool as soon as they heard the bell. They were reacting to what Pavlov defined as a conditioned stimulus: a conditioned response (the drooling) was triggered because the dogs linked the new stimulus to positive expectations (the food).

What does all this have to do with the bladder and the urge to go when you're standing at your front door?

Well, people whose bladders always get active just before they get to the front door are behaving like Pavlov's dogs. Only here, it's as if the bladder is drooling. By going to the toilet as soon as their door has been unlocked, they develop a conditioned stimulus over time. So now every time sufferers are standing at their front door rummaging around for their key, the stimulus kicks in, resulting in the learned response: the need to pee really urgently.

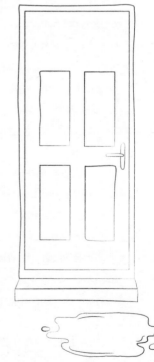

But you know the really good thing about learned behavioural patterns? You can unlearn them. In Pavlovian theory, this is called extinction. It happens when something extra or different is associated with the existing stimulus. Through this, the previous reflex is overridden and invalidated.

What does this mean for bladders that kick into action at the front door? Don't rush straight to the toilet. Go into the kitchen first, perhaps, or into your bedroom. This way, you will break the habit of this conditioned behaviour over time. If you suffer from latchkey incontinence, it's important not to give in to the urge to go immediately. Toilet training can really help you here: delay going to the toilet by a little longer each time, and record it in a bladder diary (for tips on how to resist the urge when you're out, see pages 103–7).

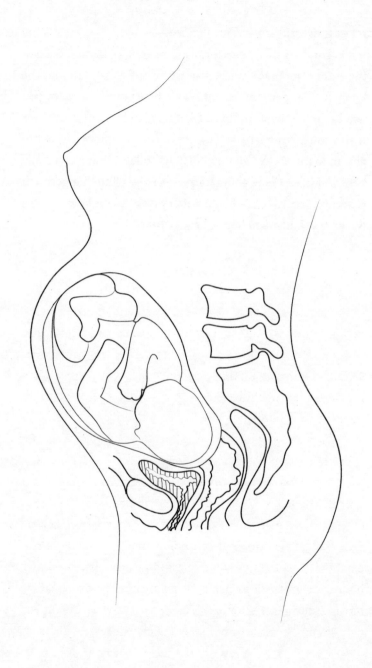

6.

You're Pregnant? What Your Bladder Has to Say About It

A tiny being has made itself comfortable in your uterus. At first, it slumbers away, demurely keeping a low profile; as time goes on, though, it grows bigger and bigger and needs more space. It's only natural that your bladder will have some opinions about having to share the abdomen with a room-mate for the next nine months.

When we get pregnant, our hormones throw a big baby shower. To stop it from getting boring, they invite lots of other noisy guests, so the number of hormones in our bodies quickly increases. VIPs at this party include the hormones oestrogen, progesterone, a hormone that goes by the name of HCG (human chorionic gonadotropin) and the love hormone, oxytocin. Generally speaking, these hormones ensure that our bodies are perfectly prepared for pregnancy and, later, for the birth. At the party, our hormones give us a spectacular make-over so our bodies are fit to meet the baby. The uterine lining

is readied for the implantation of the zygote, the tissue relaxes enough for the baby to grow unimpeded, our breasts adjust to their new role as milk producers, and so on and so forth.

The bladder also comes to the baby party. In the lower part of its mucous membrane, which is called the bladder trigone, countless receptors form to direct the effect of the new hormones into the designated cells. Receptors? That sounds familiar. Yes, there are already receptors in the bladder, ones that measure and communicate pee pressure. Now that these new receptors have come along, it gets quite cramped and confused down below. This overstimulation of the bladder is the exact reason pregnant women often start needing to go to the toilet more often right from the very beginning.

Along with this, our organs are supplied with more blood during pregnancy. And what do organs do when they receive a stronger supply of blood? That's exactly right: they run at full speed. The kidneys produce more urine and the bladder needs to be emptied more often. Of course, this is really annoying at first, but it has its advantages. Emptying the bladder more often means it is always well rinsed out, which protects it from possible assailants like bacteria that might otherwise trigger a bladder infection.

We also just drink more when we're pregnant: after all, we're drinking for two.

From around the 27th week of pregnancy, the baby starts challenging your bladder for space. After this, the organs shift and your mini-me presses on the bladder, so it has less space to fully expand and its capacity drops: the bladder must be emptied more often. Before this point, though, it is crafty and nimble enough to cleverly dodge its temporary tummy-mate.

WHY WE NEED TO GO TO THE TOILET MORE OFTEN DURING PREGNANCY

- The increase in hormones 'overloads' the bladder.
- The kidney and bladder are supplied with more blood.
- We drink more.
- From the 27th week of pregnancy, the baby presses on the bladder.

Incontinence During Pregnancy

If you leak when you're pregnant, don't fret about it. Many pregnant women have to contend with a weak bladder or incontinence during pregnancy. This is because the hormones not only make your urethra and bladder softer and looser, they also have this effect on the pelvic floor. Now of course, the pelvic floor needs to be as soft and malleable as possible for the birth, but targeted training can be done to ensure it is still strong enough to hold in urine. Tense it briefly, hold the position, and then actively relax it again. To make the exercise more intense, you can bring in the breath: tense it when you breathe out and relax it when you breathe in. You can do this exercise as often and for as long as you have time and energy. It's also great for your commute, the couch, or when you're cleaning your teeth. Importantly: don't forget to breathe, and keep your stomach and gluteal muscles relaxed.

Snoozing the Urge to Pee During Pregnancy

The urge to urinate is of course important in flushing out the bladder, so it definitely has its value. You can, however, pay attention to what you drink. For example, you can avoid typical trigger drinks such as coffee and black or green tea. Generally speaking, you should stop guzzling litres of caffeinated drinks when you're pregnant, but you know that already. You can also skip diuretic foods such as asparagus if you know that your bladder is susceptible to them (more on this on page 25).

You should also make sure you don't strain your pelvic floor too much. It already has enough to do with the new guest in your stomach. Overly long, fast walks, heavy lifting, and super-strenuous stretching exercises are now off limits. What does help the pelvic floor is good posture. Be careful not to arch your back too much or to hunch over. By working on your posture, your back muscles are trained incidentally, and they then support and relieve the pelvic floor. Breathing can help you here, too: before you lift something, whenever you stand up and so on, you should 'activate' your pelvic floor by breathing out and tensing it at the same time.

TIPS FOR CONTROLLING THE URGE TO URINATE DURING PREGNANCY

- Abstain from trigger drinks like coffee or black tea.
- Avoid diuretic foods such as asparagus.
- Talk to your doctor immediately about anything you're unsure of.
- Do gentle strengthening and preventative pelvic floor exercises.
- Wear panty liners or pads if necessary.
- Make sure you have a straight and healthy posture.
- Avoid strenuous activities.

UTIs During Pregnancy

Pregnancy makes you particularly susceptible to UTIs. You can blame the pregnancy hormone progesterone, which relaxes the muscles, makes the bladder wall softer and expands the urethra. It's a real invitation for bacteria. So please pay close attention to changes in your urination or the pee itself. As you already know: watch out for pain just before and while passing water and for any possible blood in your urine.

If you notice signs of a UTI, please go and see a doctor. If the bacteria travel farther up and cause a kidney infection, it can lead to premature labour or even a miscarriage. If you have a UTI, your doctor will do a urine culture, which will provide information about which and how much bacteria are romping around in your urine. Since not all antibiotics can be used during pregnancy, make sure you discuss the risks and side effects with your doctor and have her explain any possible alternatives.

The Baby Is Here! Now What?
The Bladder After Pregnancy

Yes, you've managed it, you've introduced a new citizen to planet Earth. Congratulations, I'm so happy for you! After you've overcome the first trials and tribulations and you're allowed to leave the hospital with your newly emerged off-spring and slowly get back into the swing of everyday life, you realise with horror that you're not entirely watertight down there. A few seconds after your bladder first gets in touch with you, urine just flows out. Or a trickle of urine unexpectedly escapes into your knickers when you cough or laugh. It might just be a few drops, or it might be the full load. Yikes. What's wrong?

Keep cool, my dear new mamas. After pregnancy, it's not at all uncommon to leak a little, or even to temporarily become completely incontinent.

Twenty to 30 per cent of all women have urinary incontinence after giving birth, and three to five per cent even suffer from faecal incontinence.[19] This is a direct consequence of giving birth; your pelvic floor has loosened during your pregnancy, and the hormones have made it soft and elastic so that labour can be as quick and painless as possible for both you and the baby. After this, it's natural for it to take some time for everything to get back to its original shape. Some women's pelvic floor muscles will have become slack or even overextended, particularly with large babies or really long and complicated births. The result: the bladder's locking mechanism has been weakened and can't hold the urine in so well anymore, resulting in stress incontinence (more on this on pages 142–50). Thankfully, these typical 'oops' moments when you laugh, cough or lift something heavy can be remedied relatively quickly and easily through targeted postnatal training and pelvic floor exercises — perhaps using vaginal cones or biofeedback (see pages 148–9) — so after a while you'll be on the move with dry pants again.

It's best for you to talk to a physiotherapist or doctor about which exercises would be most suitable for you and your pelvic floor, because there are lots of different ways to do pelvic floor training. Standing up, lying down, sitting, while doing a headstand ... the sky is practically the limit! But it's super important that you don't forget about your breathing! If you don't breathe properly while doing the exercises, your pelvic floor won't be able to contract and then relax again properly. And really: only work with your pelvic floor and leave your stomach and your butt out of it.

There are also activities such as yoga, Pilates, Qigong or gentle swimming that are perfect for the pelvic floor and postpartum recovery. They gently strengthen your core without putting too much pressure on you. And please: don't put too much physical or mental pressure on yourself. After all, you've just performed a mighty deed to become a mother — so be extra nice to yourself and your body!

7.
Bladder Care:
How to Keep Your
Bladder Healthy

It's absolutely crazy what a workaholic the bladder is, right? It's about time we showed it some appreciation! Here are a few tips on how to support your urinary equipment and even give it a little bit of love now and then.

Drink Enough

You're probably sick of hearing it, but drinking enough really is the best present you can give your bladder. You can help your bladder get healthy and stay that way by always giving it a good flushing out. This makes it harder for bacteria and germs to attach themselves to the bladder wall, where they can cause irritation or inflammation.

You should be drinking 1.5 to 2 litres of fluids every day. And you've guessed it: we're not talking about coffee or cola here. What your bladder really enjoys is water and herbal teas. If you find normal water too boring, go right ahead and make an 'infusion' by pepping it up with fruit. Just cut your favourite fruit up into smallish pieces and steep them in tap water. After about ten minutes, the water will have taken on the flavour and taste refreshingly fruity. My current favourite drink: cucumber in water. It really is deliciously refreshing.

Another thing that might help: get yourself a stylish bottle and put it on your desk next to your computer. This way you'll always have it in sight, guaranteeing you don't forget to drink regularly. Or why not play a drinking game where you're not allowed to leave your desk until the bottle is empty? I bet that from now on you'll never drink too little again, but for anyone who needs a bit more help drinking enough, there are even special phone apps that prompt you to drink at regular intervals.

Nurture your urinary system with teas

Do you often get UTIs? Or is an overactive bladder irritating you? Then try adding a range of diuretic teas to your diet! They do a really nice job of flushing out your bladder, and they can also have an anti-inflammatory or soothing effect — they're basically everything a stressed-out bladder could hope for. You can buy ready-made bladder and kidney teas at pharmacies, health food shops and drugstores, where it's a good idea to get advice on which tea mix would be best for your bladder problem. Meanwhile, pros and keen DIYers can blend teas at home themselves.

Here are two of my favourites.

To combat **recurrent UTIs**, I recommend this **bladder and kidney tea**. It flushes out the bladder really well and has an anti-inflammatory effect.

For the tea, you will need:
- 20g birch leaf
- 20g goldenrod
- 20g orthosiphon aristatus leaf (also referred to as 'cat's whiskers' because of its appearance)
- 30g bearberry leaf (but please, leave this out if you're pregnant or breastfeeding)
- 10g peppermint

To make the tea: Mix the ingredients together and put about a tablespoon in 150 ml boiling water. You can also use just two or three teaspoons of the mixture — try it out and find the amount that tastes and feels good for you. Three to five cups of this tea each day is an excellent wellness routine for your bladder.

Anyone plagued by an overactive bladder should try this **hops tea**, which soothes the bladder and the mind.

For the tea, you will need:
- 10g hops
- 10g St John's wort
- 10g buchu leaf
- 10g lemon balm
- 60g rose hip husks

To make the tea: Steep a heaped teaspoon of this mixture in a litre of boiling water — not for too long, though, or it will end up tasting bitter. Drink four or five cups of this per day, too.

Relax on the Toilet Seat

As mentioned on pages 19–21, how you sit on the toilet matters for both your pelvic floor and your bladder. It's really important that you don't strain. Just let the urine flow, and be sure to evict everything. Going to the toilet is not a race, and you can take as much time as you need and want. So don't squeeze the urine out quickly, and avoid pushing with your stomach muscles.

This is true for poo, too. If you often feel blocked up and have to really squeeze, it can eventually lead to pelvic floor damage. And if your bowels often get jammed up, it can have negative repercussions for your bladder and how it works: an overfilled rectum can really expand and press on the bladder. To get everything running smoothly in the bathroom again soon, you should take a closer look at your eating habits and perhaps speak to a nutritionist.

Appreciate the work of the entire urinary system. We should stop seeing a trip to the loo as something annoying to get over and done with as quickly as possible. Take your time, and silently thank your bladder for being such a good friend and for working — after all, this can't always be taken for granted.

Eat a Healthy, Well-Balanced Diet

You can also be kind to your bladder through your diet. Vitamins and minerals strengthen the body's general resilience. And, as you know, a good immune system is the best protection against infection and illness, so make sure you eat plenty of fruit and vegetables every day. Pumpkins, or rather, their seeds, are particularly good at soothing the bladder. You can try taking capsules from a pharmacy or chemist, or enjoy the pure, unadulterated seeds themselves each day — they're also super tasty when toasted.

Owners of sensitive bladders should avoid highly acidic foods such as citrus fruits and pineapple, because they can irritate the bladder even more. In general, just be mindful of which foods do you good (and taste nice, too, of course).

181

The Right Clothing

So … what are you wearing? I'm not trying to be suggestive, it's just that the answer is really important for your bladder and genitals. If they were to go lingerie shopping with you, only loose cotton undies would end up in your shopping bag. Knickers that are too tight or, worse yet, made of synthetic materials, can irritate the delicate tissue of the vaginal and urethral openings, triggering unpleasant infections. If you want to do your vulva and bladder a favour, switch to comfortable cotton underwear. You don't have to throw away your sexy thongs, just stop wearing them every day — save them for special occasions instead. When washing your knickers, make sure that the detergent isn't too aggressive or perfumed, as this can also cause irritation.

Jeans can sometimes be to blame for stressed lady bits and bladders, too. If they're too tight, they not only make you insanely uncomfortable and are hard to get on and off, they rub and cut into the crotch, making your genital area the perfect landing strip for bacteria and viruses.

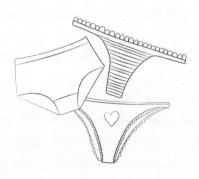

Keep Your Feet Warm

Cold feet are not only the worst enemy of cosy nights on the sofa or hours of romance under the sheets — they also make life hard for your bladder. The cold doesn't damage the bladder directly, but it does weaken the immune system. In the cold, the body switches to low-power mode to conserve energy. Less blood flows to the feet and hands, and they cool down. If we don't warm our feet up, blood flow to the body's mucosa — including of course the bladder mucosa — is reduced, weakening our immune system, which in turn means bacteria, germs and viruses can more easily infiltrate and damage our (bladder) health. To avoid this, just stash some thick, ultra-comfy woollen socks next to the sofa. Or put a hot water bottle on your feet.

Send cold feet to a watery grave with warming footbaths

Another thing that's great: warming footbaths. What I mean by 'warming' here is that the temperature keeps increasing. Here's how to do it: put water of about 30 degrees Celsius into a basin (or the bath) and stick your feet in. Roughly every two minutes, top it up with hotter water until it's reached about 40 degrees. Once that's happened, sit there for another 15 minutes or so and enjoy the feeling of the warmth slowly rising up from the bottom to the top of your body.

Doing this warms up the body and expands the blood vessels step by step. What's really cool (or rather, hot) is that this slow, incremental warming doesn't just feel heavenly, it can help with a sensitive bladder, too: it is beneficial to the immune system and keeps the mucosa strong and healthy.

Ensure Your Spine Is Supple and Healthy

All the nerves responsible for the bladder are channelled through the vertebrae of the sacrum and tail bone and then up through the spine to the brain. So naturally, the health of our back affects the health of our bladder. The spine is very sensitive, and even minor back injuries can have a big effect on the urinary tract. It's important to have a strong back, but strengthening exercises must be done properly, or they can do more harm than good. If you're unsure about how to approach back training, or are already in pain, please don't hesitate to consult a specialist.

If your job means you spend most of the day sitting (welcome to the club), it can really impact your back. Make sure you arrange your desk, including your computer screen, so you're not sitting hunched over or with a swayback. Poor posture can squeeze your organs together or reduce the tension in the pelvic floor. So, from now on, ladies: sit up straight!

If you have a desk job, it's also good for your back to stand up, stretch, and walk around a bit every now and again — maybe even to go to the toilet!

Say Yes to Intimate Care, But Please Don't Go Overboard

Yeah, yeah, advertisements and some women's magazines always want to make us believe that our private parts are dirty and smelly, and need to be purified with a wide range of cleansing products. I hope you already know that this is mostly rubbish. Our vulva is quite a clean organ and, unlike the penis, has a self-cleaning mechanism. It simply washes away dead flakes of skin, cells, bacteria and so on, in the clear or white discharge that we find in our knickers every day. This discharge is mainly made of lactic acid bacteria, which keep the pH value of our vaginas nice and acidic. Bacteria and viruses don't like acidic environments one bit, so they quickly vanish instead of multiplying. Our genital area also has a mucosal surface that acts like a kind of slide, making it harder for bacteria and viruses to settle down and get comfortable.

So, if we go crazy cleaning our genital area, we also scrub away the important sentries whose job it is to keep bacteria and viruses at bay. The result: our genitals are as good as defenceless. This is why you shouldn't go overboard in how often you wash your intimate area. Mild shower gels and luke-warm, fresh water are most suitable. It can also be sensible to have an extra towel for your intimate area that you change regularly. After all, you don't dry your face with the same towel that you just rubbed your feet down with, right?

Pelvic Floor Exercises

As you should know by now, when it comes to supporting your bladder through all possible situations in life, nothing beats a strong and healthy pelvic floor. And this fact shouldn't just appear on your radar once your pelvic floor falters and stops working as you'd like it to. Prevention here really is better than a cure, so it's advisable to integrate pelvic floor exercises into your everyday life. The good thing: you don't need a fitness studio or any fancy equipment to exercise your pelvic floor or to actively relax it. You just need yourself and your pelvic floor. See pages 118–19 for more advice on pelvic floor exercises.

Alongside targeted exercises that actively strengthen the pelvic floor, you can also make sure that you look after it in everyday life. By and large, this is about protecting the pelvic floor and avoiding putting too much pressure on it. This way, it will stay strong and support you and your bladder for a long time to come.

Getting up properly

Your alarm is going off and you have to drag yourself out of bed even though you're still half asleep. But what if I told you that HOW you get up is as important as when? Your pelvic floor doesn't like it one bit when you heave yourself straight up from lying supine like Countess Dracula (and, by the way, neither does your back). It's much healthier to first roll onto your side, support yourself and then use your arms to push your entire upper body upright. Yes, this takes some getting used to at the start, and of course you'll forget to do it sometimes, but believe me, after a while you won't even remember that you used to get out of bed any other way.

Heavy lifting

Whether you're moving house, cleaning up, or looking after children or pets, we all have to do heavy lifting sometimes. There's certainly a risk here of acute muscle soreness or pain in the lower back, but the pelvic floor can also be affected. This happens when you get the strength to do the lifting not from your legs but from your back. In doing this, you put enormous pressure on the pelvic floor — not to mention your back. Your entire body will thank you if from now on you bend your knees and then use your leg muscles to lift the object, child or pet, keeping your back straight. Another important thing that is sadly often forgotten: breathing properly. To foster teamwork between the abdominal, back and pelvic floor muscles, it's important that you keep breathing normally and don't hold your breath.

What to do when you've got a cold

Are you fighting a running battle with a nasty cold and coughing and sneezing non-stop? Well, first of all, get well soon. Over time, this constant sneezing and coughing doesn't just get on your nerves, it can also get to your pelvic floor. Because with every cough and every *atishoo*! your body builds up pressure, which it piles onto the pelvic floor. To reduce this pressure next time you sneeze or cough, give the following a try: instead of facing straight on, turn the upper part of your body to the side. And of course, cough or sneeze into your elbow and not into your hands.

Work *with* your bladder, not against it

In my opinion, this is one of the most important points. Be kind to your bladder and accept its needs. Stop working against it and aggravating it. If you know that it reacts strongly to coffee or gin and tonic and will send you to the toilet more often if you have them, drink something else. If you do drink them, don't complain when you have to pee every twenty minutes. Stop having angry conversations with your bladder when it gets in touch with you at a super-inconvenient moment yet again.

Make friends with your bladder and its co-workers, such as the kidneys, pelvic floor and so on. Try to get to know them and understand why they react like they do. Look after them by having enough to drink, eating well and exercising, but also by resting and relaxing.

Pay Attention to Your Fitness

Exercise is good for the body and the mind. Your bladder also loves it when you move regularly and do activities that promote circulation and strengthen the core and pelvic floor, such as yoga, Pilates, bike riding and swimming.

If the bladder and pelvic floor are healthy and invigorated, you can run riot with sports and fitness: anything is permitted as long as it does you good. A healthy body weight — in other words, having neither too much nor too little on your ribs — is good not just for your bladder but for your physical and emotional health, too.

If, on the other hand, you leak a little now and then, and your pelvic floor isn't exactly in top form at the moment, you should avoid physical activities that are too wild. And by this I don't just mean extreme sports like climbing, deep-sea diving, or base-jumping; any exercise in which you have to jump, move quickly or stop suddenly puts a lot of pressure on your bladder and pelvic floor. This list of no-go sports includes squash, skipping, tennis, football and so on. Before you go ahead with these, do a quick check on how your bladder is doing.

The Bladder as the Mirror of the Soul

As we've learned, there are some urological conditions that cannot always be explained organically, such as overactive bladder, which is the most frequent somatic symptom disorder amongst women.[20] According to some experts, 'at least 15 per cent of women with cystitis symptoms have no organic basis for their complaints'.[21] This is a really remarkable number, right? Sayings such as 'I was so scared I nearly wet myself' or being 'pissed off' obviously don't come from nowhere. We also need to go to the toilet more often when we're nervous or under pressure. But why?

Why we need to go to the toilet more often when we're nervous

The autonomic nervous system is responsible for our sprint to the toilet before a job interview or an important meeting. It controls our bladder function. As we learned on pages 42–4 about paruresis, the autonomic nervous system includes the sympathetic and the parasympathetic nervous systems. The sympathetic nervous system ensures that our bladder can collect urine. It keeps the sphincter nice and tight and the bladder muscle nice and relaxed so that the bladder can expand. The parasympathetic, on the other hand, is responsible for dispatch, so it brings about the exact opposite: the bladder muscle is activated and contracts, the sphincter loosens its grip and the urine can flow out.

Now, having an important test, giving a speech or waiting outside a job interview makes us nervous and stressed. Our heart rate increases and we start to sweat and tremble. To balance this out, the parasympathetic nervous system is activated. It is supposed to power down and calm the system, *i.e.* our body. And how does it do this? Amongst other things, by dilating the blood vessels in the skin, stimulating the production of sweat and even increasing bladder activity.

Perhaps you can imagine this as a kind of pressure adjustment system designed to relieve the body. So, if you have sweaty hands before a job interview, discover red patches on your face, décolletage or throat, or need to go to the toilet every ten minutes, your body isn't doing it to annoy you or make you even more stressed; it's entirely the opposite. It's trying to help you relax.

Our bodies also react to situations of shock or panic with a sudden urge to urinate or even actual leakage. This is a function of evolution. When our ancestors were standing in front of a snarling sabre-toothed tiger, the body immediately switched into flight mode. To make itself as prepared for this as possible, it cast off any dead weight it was carrying: faeces and urine had to be evacuated as quickly as possible.

So, when our nerves are raw, our bladder, pelvic floor and sphincter muscles are no longer controlled properly. They get confused and stop working with each other, as if they've lost their team spirit. If the nervous strain continues for a long time, this can manifest in bladder dysfunction.

Another sign that a bladder problem is of a psychosomatic nature is that it only happens when you're awake. If, for example, your bladder sends you to the toilet super urgently every twenty minutes during the day, but usually lets you sleep peacefully through the night, it's likely that your problem is psychosomatic.

Problems with the bladder can also occur as secondary symptoms of depression, sexual dysfunction or as the expression of various anxiety disorders such as claustrophobia; unfortunately, they can also make these conditions worse.

It's impossible to generalise about which psychological sensations trigger and influence the bladder the most. Patients with psychosomatic bladder dysfunction are a very heterogeneous group with different problems.

The urinary bladder type according to Barral

According to the French osteopath Jean-Pierre Barral, each organ is associated with specific emotions, which, in the long run, can make it weakened, irritated and sick. To understand this better, he divides people into different organ types and describes what these types represent and why certain patterns of behaviour and thought often lead to problems with particular organs. Along with the heart type, the genital type and the breast type, there is of course also a urinary bladder type.

For Barral, the origins of bladder problems are in childhood education. If your upbringing was really strict with lots of prohibitions, punishments and discipline, this can express itself in adulthood as uncontrollable anxiety and feelings of guilt. Those affected have trouble trusting and accepting their body and its functions. Such internal tension can have a big influence on the bladder and how it works.

In addition to this, the bladder is the organ that represents control for Barral. He thus sees people with bladder problems as control freaks — not towards others, but towards themselves. Everything has to be perfect; they can't let themselves go, or put things off for a while. Bladder-type individuals are really scared of being judged by others and of appearing in a bad light. For this reason, they often try to make everything right and not be a nuisance. For the bladder type, attracting negative attention is out of the question. At some point, this permanent pressure and tension must come out of the body somewhere, and it's not uncommon for it to be through the bladder.

What you can do about psychosomatic bladder problems

To improve or alleviate the symptoms in the long run, targeted psychotherapy is important. Through this, patients should come to recognise how and why their physical symptoms are linked to the psyche. Why is your bladder acting up when the problem is somewhere else? You can also get to the bottom of old or perhaps even unconscious patterns of behaviour. They can be worked on in order to ease potential anxiety or obsessive-compulsive disorders. Relaxation exercises and stress management strategies are important for this, and can be integrated into everyday life.

8.

Funny Facts About
the Bladder

And now, to round things off, something for the nerds amongst us. This chapter is for anyone who likes trivia. Because believe me, the bladder and its functions can be a real icebreaker at parties. And not just when you're waiting in the much-too-long line for the bathroom, yet again.

A Historical Digression

In the past, people knew that you could do a thing or two with the bladder. Animal bladders were not just thoughtlessly thrown away as slaughterhouse waste, but instead turned into a wide range of implements. After all, the bladder wall is naturally light yet still quite robust. This means it's perfect for making tools for transportation, such as sacks and bags.

And bladders weren't always just used for practical purposes! Pigs' bladders were often used as toys. Children would blow them up or fill them with water to make balloons or water bombs, and they were particularly good for football games. King Henry VIII of England — you know, the guy with the six wives — apparently loved kicking pig's bladder footballs around the pitch at court.

Adults also used them to have a different kind of fun. That's right: the first condoms were made of animal bladders. Legend has it that King Minos of Crete was the first real condom user. He thought that the first semen he ejaculated would be lethal to his beloved wife Pasiphae, so he protected her by using a condom made of a goat's bladder. Later in antiquity, Greeks used animal bladder condoms regularly — not to prevent pregnancy, but to protect against venereal diseases.

And what would the Inuits, the original inhabitants of Alaska, think of this? They believe that the soul of an animal is in its bladder, so using an animal bladder as a condom replacement would be out of the question. To properly celebrate slain animals — mainly seals — and to thank them for their bounty, every year in the middle of winter Inuits would celebrate Nakaciuq or Nakaciuryaraq, which translates as Bladder Festival. The bladders of all the slain animals were collected, dried, inflated and painted beautifully. These redesigned animal bladders were then given back to the sea to honour each of the animals. This was supposed to show respect to the former bladder owners and pray for a good hunting season to come. The Inuits also believed that new animals would arise from the bladders, thus maintaining the natural balance of things.

Can Someone Else's Urine Kill You?

Tell me you've never asked this question. Maybe it was during a high-adrenaline film in which the protagonists couldn't find anything to drink and had to come up with alternatives, or during a juvenile game of 'truth or dare'. The fact that you can drink your own urine and that some people use it therapeutically is nothing new (more on this on page 50). But what about other people's urine? Could you drink that in an emergency situation, too? Or is it poisonous because it's not your own?

Well, as long as the urine donor is healthy and the urine is fresh, you can drink it without worrying. 'Without worrying' doesn't mean that it's necessarily nice, though, or that I would recommend it. It's also important that the person chosen as a urine donor has had enough to drink beforehand, that is to say, that their fluid reservoir is full enough. If this isn't the case, the urine will just be made up of waste products that the body wants to get rid of.

Give me your juice and I'll give you mine

There are even people who get turned on by other people's urine. Urophilia describes the sexual predilection for peeing on someone or having him or her pee on you. When this foreign urine lands not on the skin but in the mouth and is then happily swallowed down, it's called urophagia. To make it sound a bit more fun and less like a serious disease, some people call it warm champagne, a golden shower, water sports or simply pee play.

The Strange Pee-haviour of Animals

What does an elephant have in common with a cat? Or a raccoon? Or a hyena? No, they're not the same size and nor do they have the same build — not even close. Despite this, they all take the same amount of time to tinkle. About twenty-one seconds, to be precise. And this despite the fact that the bladder of an elephant can hold about 18 litres of urine, while a cat can only carry a dainty 5 millilitres. Crazy, huh?

This was discovered by a research team in Atlanta. They were investigating the toilet habits of zoo animals that weighed over 5 kilograms: how they pee, when and how much. What came out was that they all emptied their bladders in 21 seconds, and all roughly five to six times per day. This probably has to do with the length of the urethra and gravity. Only animals of a certain size — those weighing over 3 kilos — are capable of forming a urinary stream. Smaller and lighter animals do not have this privilege; they just gently drip when they want to empty their bladders.

Sloths: every trip to the toilet could be their last

We all know sloths are neither the most active nor the quickest of animals. But did you know that they are so lazy they only empty their bladder and bowel once a week at most? Well, not lazy; it's just that their metabolism is super slow. When their bladder or bowel does strike, these cute animals spare no effort in getting down to it: they leave their comfortable tree nest and scramble in the direction of the ground. And we all know how long this takes for sloths, and how arduous it is. Once they get down, they dig a pit to do their business in. They are slothful, not slovenly, after all.

The tragic thing about this whole business? While the sloths are concentrating entirely on answering the call of nature, they are very easy prey for their natural enemies. This is why roughly half of all sloths die while going to the toilet.

Seriously though, why don't they just let their loads drop or gurgle from the treetops? Researchers have discovered that pee and poo are not the only reasons sloths embark on this dangerous journey downwards. It's to do with the green algae that grow on sloth skin. These algae don't just provide the animals with a nice snack, they also camouflage them from birds of prey. To keep the algae magnificently and beautifully green, little moths live in sloths' fur and lovingly look after the algal growth. These moths lay their eggs in sloth poo. When the baby moths hatch, they feed on the sloth faeces and then as adults fly up into the tree, heading towards the green fur of the sloths. In their luggage, they take along all kinds of sloth excretions that make great fertiliser for the algae. So, you see, it's a thoroughly fruitful, if risky, menagerie à trois.

How to make a horse pee

Imagine this: when someone whistles in your ear, you need to go to the toilet. Funny, right? Yeah, unless you find yourself at a sports event with lots of caterwauling and whistling — then it would be rather uncomfortable. This is exactly what it's like for horses, though: they start to tinkle as soon as they hear a whistle. It's not exactly clear why this is so, but it's assumed that the whistling somehow calms the animals down, so they release their tension and open their urethral sphincters. This technique is used before horse races so that the horses don't gallop off with full bladders. It's also handy for conducting anti-doping tests on their urine.

An Easter Egg with a Difference

Chinese spring eggs — sounds good, right? How they are cooked takes some getting used to, though. They're not cooked in normal water; they're cooked — you might've guessed it — in urine. Specifically, children's urine.

In the Chinese city of Dongyang, what's called a Virgin Egg or Boy Egg is considered a delicacy, and the city has even designated it part of the local intangible cultural heritage. Alright, alright. Apparently the eggs are not just really delicious but really healthy, too. According to traditional wisdom, they are supposed to help with joint pain, fever and fatigue.

For the eggs to develop their full effect, only pee from ten-year-old boys should be used. It's unknown why girls' urine isn't suitable. The urine is collected voluntarily in schools, where there are tubs that boys are supposed to pee into. The eggs are washed and then boiled in hot urine for about an hour. With the shells on. On the second pass, the shells are cracked so that the eggs can really soak up the children's pee. Delicious! At the end of the urine bath, the eggs are hard-boiled and have taken on a golden-brown colour. They are said to have a slightly spicy flavour.

Pee Power:
'Go Ahead and Pee, I Need Some Light'

Can you do good for yourself, your fellow humans and the environment while tinkling? Yes, it is possible. You see, researchers from the University of Bristol have managed to create electricity from urine. This works because, along with water, urine also contains carbohydrates that are eaten by bacteria, thus setting off a metabolic process. So for once it's good that our urine is a breeding ground for bacteria.

To make electricity from urine, you need a ceramic cylinder known as a microbial fuel cell that has various electrodes attached to it on both the inside and the outside. Now, when the carbohydrates in the urine are broken down by the bacteria, positive and negative particles are generated. The negative particles are transported into the fuel cell by a wire, where they create an excess of negativity. Because negatives attract positives and vice versa, the positive particles now move to the inside of the fuel cell. And voilà, energy is released and electricity is generated.

Pee power isn't really worth it on a large scale yet, because you need quite a lot of these fuel cells, and the cost of materials, construction and transport are still too high. But British scientists have already shown how great and how efficient electricity from pee can be: in 2015, they installed a fuel plant with about 400 fuel cells at the Glastonbury music festival. Pee from festivalgoers powered the light in the toilets at night. Electricity is also being made from urine in Nairobi, where there is a girls' school with a pee plant that powers enough light that the girls can pass water safely, even at night.

Insane in the Mucous Membrane: The World's Largest Urinary Stone

Normally, urinary stones are just a few millimitres wide and weigh practically nothing. But not always. The largest urinary stone in the world was discovered in Hungary, and it made all the urinary stones that came before it seem like sand. Its fighting weight: a hefty 1.125 kilograms, and it was as big as a honeydew melon. The stone was discovered in a routine examination, since bizarrely the bearer of the stone had experienced neither pain nor trouble urinating.

To Pee or Not to Pee, and Other (Un)Necessary Knowledge

Which toilet should I use? What should I do at the cinema when I really need to go? And can women pee standing up? Finally, the answers to all your most pressing questions.

The pee rule for the perfect trip to the toilet

To all you men out there: there is a formula to stop your pee from going sideways. Finally, no more splattered toilet bowls and trousers. And boys, you can master this formula even if you're bad at maths. This anti-splashback tactic was invented, intensively researched and approved by a team at the Brigham Young University in Utah.[22]

If you want to pee sprinkle-free from now on, it's not just where you aim, but the angle at which you hit the target that's important. Aiming perpendicularly so that the urine hits the wall or water at a 90-degree angle is no good for splashback. When it comes to the angle at which the urine shoots out — the dangle angle, if you will — the smaller the better: men need to narrow the 'angle of attack'.

The researchers also discovered that men should stand as close to the toilet as possible. This way, nothing can go askew if the stream breaks off early or there are pauses. By the way, the stream usually breaks up into droplets when it is about 15 to 18 centimetres away from the urethra. So, if men stand farther from the bowl, *e.g.* 20 centimetres away, part of the stream will of course miss its target, and the multiple droplets will also cause more flashpoints — or rather, splash-points.

The physicists also discovered something that men might not like to hear: for the ultimate in splash-free urination, they should just sit down.

So men: whichever formula you choose to solve the problem, you have no excuse for missing the target now.

Ladies, stand up! How women can piss on their feet, too

We women can also pee standing up. This is naturally super practical when you're out and would really like to just go quickly behind a bush. Thankfully, there are now various tools to help us pee standing up, so the annoyance of having to squat and pray both that no one discovers you and that you don't pee on yourself is now a thing of the past.

A female urination device such as a Urinella or a Shewee can make this possible. It looks like a little funnel: the urine goes in at the top and drains down and out through the spout.

To stop anything from missing the mark, the top is placed directly on the urethral opening between the labia, and you hold the end of the Urinella out and — importantly — away from your clothes. If you do everything properly, your stance should be like a man's when he pees. And yes, you can also fool about with the Urinella by conducting the stream to make funny patterns with your wee.

Urinary aids are usually made of silicone, so you can use them over and over again, but there are also products made of cardboard that you fold up and throw away after urinating.

There are even pissoirs for women. Sure we can't stand completely upright at them like men, but we don't have to sit down, either. The gentle half squat, called the 'skiing position' by pros, is the perfect posture here. This way, our thighs and buttocks don't have to touch any gross, cold toilet seats, which is of course more hygienic as well as being quicker. Squat toilets are common in lots of countries all over the world, for example in India, the Islamic world and Southern Europe, although unfortunately you rarely come across them in the Anglosphere or in my home country of Germany.

Why we need to go to the toilet so often under water

You're relaxing in the bathtub, lathered in bath oil, candles lit and chillout music on, and you're totally blissed out, when suddenly, your bladder calls. Oh man, why now? The so-called Gauer-Henry reflex is to blame for this. Our blood flow changes under water. On land, gravity pulls our blood down to the legs and abdomen. When we're in water, the veins are compressed so our blood now needs to take different routes, and more of it is piped into the middle of the body, *i.e.* the stomach and chest areas. As a result, stretch receptors in the atria of the heart are activated, and pressure increases. To balance out this pressure, the heart and lungs send nervous signals to the kidneys, triggering an increased urge to urinate in order to create equilibrium.

When the stretch receptors are activated, the production of antidiuretic hormone (ADH) is also put on hold. ADH ensures that our body has enough water in reserve, so that we don't dehydrate. Now that our body has stopped releasing enough ADH, more water in the kidneys is turned into urine. This is why we need to go to the toilet.

Incidentally, this doesn't happen to astronauts in space; all the weightlessness does there is cause the message that the bladder needs emptying to be relayed much too late. Because of this, NASA has developed special nappy-like shorts for take-off, landing and hanging out in space.

Tinkling at the cinema: how to stop missing exciting scenes

It's happened to practically everyone: you're sitting in the cinema, totally absorbed in the film, when suddenly you need to go to the toilet. You now have two options: either you quickly duck to the loo, meaning you'll miss crucial minutes of the film; or you stay where you are and sit in torment until the end of the movie. Hmm, neither option is exactly appealing.

The good news is that there's an app that can help you in exactly these situations. It's called 'RunPee' and it shows you exactly when the right moment for a toilet break will be coming up — in other words, it tells you when there are boring, unnecessary scenes in the film. It also tells you how long you have for your toilet break and when the next possible opportunity to relieve yourself will be. Cool, huh?

How to always nab the cleanest toilet in a public loo

And finally, another super-important tip. Which cubicle in a public toilet do you think is likely to be the cleanest: the first one? The very last? Or do you perhaps aim for the middle?

And, the award for the cleanest toilet goes to — drum roll — the first toilet cubicle! Congratulations! The reason for this is that the first toilet is just too close to the entrance for many people. They want to enjoy some peace and privacy when they're taking care of business. Some people are so shy that, if they are able to go in a public toilet at all, they can only use the last cubicle. And this is exactly why the first toilet is used the least, thus making it cleaner. So remember that for your next trip to the toilet.

Thanks

Wow, I've done it! What a wild ride. But hey, now we've really arrived at the finish line, and I'm infinitely proud.

First up, I would like to thank my two sisters.

Annette, without you I never would have written this book. You were there when my irritable bladder first made itself known, and you gave it a face with your fantastic illustrations. As my twin sister, you're also my kindred spirit, and you've built me up and supported me through the crappiest situations. Tine. My big sister. If you hadn't always pushed me and given me courage, I might not have even dared to write my own book. You have also become a mother, and given me and Annette the sweetest nephews anyone could wish for.

Mama, Papa, Katja, Basti, and of course my little Jonathan. Thanks for rescuing me from all kinds of situations and always being there for me. I love you all so much! And naturally my godfathers, Horst and Klaus, and the entire Bulla clan. I'm sorry that I can't mention you all by name, but please know that I'm speaking to all of you.

I would also like to thank my cuddly co-author: Domi, my little Spanish street cat. You have comforted me and stayed up with me while I worked through the nights. (This is something very close to my heart: #adoptdontshop. There are so many animals looking for safe, loving homes.)

From a medical perspective, my thanks go to Dr Wolfgang Bühmann (urological specialist), who was always there when I had medical questions. Just like Prof. Dr Daniela Schultz-Lampel (director of the clinic and principal consultant for the Kontinenzzentrum Südwest at the Schwarzwald-Baar-Klinikum and member as an expert adviser to the Deutsche Kontinenz Gesellschaft); and Dr Jessica Kruse. Thank you for all the subject-specific support. Many thanks to the Deutsche Kontinenz Gesellschaft (especially Julia Ehlers) and the German Urology Society for your fantastic help putting me in touch with experts.

Dear Ulrike and the Hanserblau team. Thanks for believing in this idea and in me, and for placing so much trust in me. Thank you to Daniel Mursa for acting as my agent. Oh, and of course Volker Wittkamp. Without this funny chain and your efforts, the book would never have come about as it did. Sometimes life is peppered with funny circumstances.

And, of course, my friends. Thanks for being proud of me and believing in me. Without you, I would have despaired sometimes, and I wouldn't have been able to enjoy the good times, either. Here is a short but very sweet list (in alphabetical order): Andrea, André, Chris, Christina, Claudia, Daniel, Derk, Flo, Frodo, my femininINNEN crew, especially Katja for the great photos, my Goldies, Jassi (my thoughts are often with you), Nadine, Simone, Stefanie and Thomas (my dear Burzans), Tim and Tobi.

And, last but not least, I would also like to thank you, my dear bladder. It would be a lie to write 'thank you for being who you are'. I would rather write 'challenges make you stronger'. Otherwise I would not be sitting here writing these lines.

Oh yeah, and even though some people might think it sounds a little strange, I would also like to thank myself and say how proud I am of me. For overcoming embarrassment, practising self-love and accepting and loving who I am. Thanks, Birgit!

We are all unique and it is really important that we accept ourselves — love ourselves, even — for who we are. We all do this much too little. So let's all support each other, with all our flaws. And write books about them. I'm excited, and I'm looking forward to whatever comes next.

Endnotes

1   Stephen W. Leslie, Hussain Sajjad, and Patrick B. Murphy, 'Renal Calculi', *The National Center for Biotechnology Information* (2021). https://www.ncbi.nlm.nih.gov/books/NBK442014/

2   Finlay Macneil and Simon Bariol, 'Urinary Stone Disease: assessment and management', *Australian Family Physician 40*(10) (2011). https://www.racgp.org.au/download/documents/AFP/2011/October/201110macneil.pdf

3   'Urinary Tract Infections', *University of California San Francisco*. https://www.ucsfhealth.org/conditions/urinary-tract-infections

4   Ildikó Gágyor, Jutta Bleidorn, Michael M. Kochen, Guido Schmeimann, Karl Wegscheider, and Eva Hummers-Pradier, 'Ibuprofen versus fosfomycin for uncomplicated urinary tract infection in women: randomised controlled trial', *BMJ* (2015). http://www.bmj.com/content/351/bmj.h6544

5   Nora Zamichow, 'Diaphrams linked to urinary infections', *The Washington Post* (1985). https://www.washingtonpost.com/archive/lifestyle/wellness/1985/07/17/diaphragms-linked-to-urinary-infections/ce50245f-f3c8-47b7-ae9f-7b9d504707b2/

6   'Industrielle Tierhaltung braucht Antibiotika – und erhöht das Risiko resistenter Bakterien', *BUND*. https://www.bund.net/massentierhaltung/antibiotika/

7   Kalpana Gupta, 'Urinary Tract Infections (UTI): Pyelonephritis/Complicated', *Infectious Disease Advisor* (2017). https://www.infectiousdiseaseadvisor.com/home/decision-support-in-medicine/infectious-diseases/urinary-tract-infections-uti-pyelonephritis-complicated/

8   Ines Ehmer, MD, Blasenentzündung und interstitielle Zystitis: Test – Therapie – Schmerzbekämpfung, Zuckschwerdt Verlag (2019).

9   Ibid.

10 'New data highlighting considerable burden of Nocturia', *Ferring*

Pharmaceuticals (2013). https://www.ferring.com/new-data-highlight-considerable-burden-of-nocturia/

11 Marco Hafner, Wendy M. Troxel, Michael Whitmore, and Christian Van Stolk, 'How trips to the bathroom at night affect health and productivity', *RAND* (2019). https://www.rand.org/randeurope/research/projects/nocturia-effects-on-health-and-productivity.html

12 'Endometriosis', *World Health Organization* (2021). https://www.who.int/news-room/fact-sheets/detail/endometriosis#:~:text=Endometriosis%20affects%20roughly%2010%25%20(190,and%20girls%20globally%20(2

13 Dr med. Dirk Manski, 'Endometriose mit Befall von Harnleiter oder Harnblase', *Urologielehrbuch.de*. https://www.urologielehrbuch.de/endometriose.html

14 Ian Milsom, 'How big is the problem? Incontinence in numbers', Gothenburg Continence Research Center (2018). https://www.gfiforum.com/Upload/43b34997-7408-4fa6-9547-72488e668060/I%20Milsom%20-%20Incontinence%20in%20numbers.pdf

15 Victor W. Nitti, MD, 'the Prevalence of Urinary Incontinence', *Reviews in Urology 3* (Suppl 1) (2001). https://www.ncbi.nlm.nih.gov/pmc/articles/PMC1476070/#B1

16 'Incontinence: it's time to break the silence', *Hartmann* (2019). https://www.hartmann.info/en-corp/articles/d/8/incontinence-time-to-break-the-silence

17 H. H. Thyssen, L. Clevin, S. Olesen, and G. Lose, 'Urinary Incontinence in Elite Female Athletes and Dancers', *International Urogynecology Journal 13* (2002). https://doi.org/10.1007/s001920200003

18 K. Eliasson, and T. Larsson, and E. Mattsson, 'Prevalence of stress incontinence in nulliparous elite trampolinists', *Scandinavian Journal of Medicine and Science in Sports* (2002). https://pubmed.ncbi.nlm.nih.gov/12121428/

19 Robert Bublak, 'Inkontinenz nach der Geburt bleibt meist', *Ärzte Zeitung* (2013). https://www.aerztezeitung.de/Medizin/Inkontinenz-nach-der-Geburt-bleibt-meist-275594.html

20 Christoph Hammes, Elmar Heinrich, Tobias Lingenfelder, and Christine Cotic, *Basics Urologie 4th edition*, Elsevier (2019).

21 Alfred Auerback and Donald R. Smith, 'Psychosomatic Problems in Urology', *The Western Journal of Medicine*, California Medicine, (1952). https://www.ncbi.nlm.nih.gov/pmc/articles/PMC1521180/

22 James Morgan, 'Physicists probe urination "splashback" problem', *BBC* (2013). https://www.bbc.com/news/science-environment-24820279

References

Jean-Pierre Barral: *Die Botschaften unseres Körpers: Ganzheitliche Gesundheit ohne Medikamente*, Irisana, 2013.

Tim Boltz and Jule Gölsdorf: *Harn aber herzlich: Alles über ein dringendes Bedürfnis*, Piper, 2015.

Nina Brochmann and Ellen Støkken Dahl: *Viva la Vagina: Alles über das weibliche Geschlecht*, S. Fischer, 2017.

Dr med. Ines Ehmer: *Patientenratgeber 2019 Blasenentzündung und interstitielle Zystitis: Test – Therapie – Schmerzbekämpfung*, Zuckschwerdt Verlag, 2019.

Ines Ehmer and Michael Herbert: *Probleme im Intimbereich ... damit müssen Sie nicht leben*, 4. Auflage, Zuckschwerdt Verlag, 2016.

Dr Andrea Flemmer: *Blasenprobleme natürlich behandeln: So helfen Heilpflanzen bei Blasenschwäche und Blasenentzündung*, Humboldt, 2015.

Christoph Hammes, Elmar Heinrich and Tobias Lingenfelder: *BASICS Urologie*, 4. Auflage, Elsevier, 2019.

Richard Hautmann and Jürgen E. Gschwend: *Urologie*, 5., aktualisierte Auflage, Springer Lehrbuch, 2014.

Dr Christoph Pies: *Was passiert beim Urologen?* Herbig, 2017.

Dr med. André Reitz: *Gesunde und starke Blase: Erfolgreiche Behandlung von Blasenstörungen und Inkontinenz*, S. Hirzel Verlag, 2010.

Gisela Schön and Marco Seltenreich: *Inkontinenz: Ein mutmachender Ratgeber für Betroffene, Angehörige und Pflegende*, Maudrich, 2011.

R. Tanzberger, A. Kuhn, G. Möbs, U. Baumgartner, M. Daufratshofer and A. Kress: *Der Beckenboden – Funktion, Anpassung und Therapie: Das Tanzberger Konzept*, 4. Auflage, Urban & Fischer Verlag/Elsevier, 2019.

Volker Wittkamp: *Fit im Schritt: Wissenswertes vom Urologen*, Piper, 2018.

Arbeitskreis Blasenfunktionsstörungen, G. Primus (Vorsitzender), H.

Heidler: *Leitlinien Blasenfunktionsstörungen*, Journal für Urologie und Urogynäkologie 4/2003.

AWMF: *Leitlinien für Diagnostik und Therapie in der Neurologie: Diagnostik und Therapie von neurogenen Blasenstörungen*, Entwicklungsstufe: S1, Federführed: Prof. Dr W. H. Jost, Freiburg, 2015.

Interdisziplinäre S3 Leitlinie Epidemiologie, Diagnostik, Therapie, Prävention und Management unkomplizierter, bakterieller, ambulant erworbener Harnwegsinfektionen bei erwachsenen Patienten, Aktualisierung, 2017.

Robert Koch-Institut Statistisches Bundesamt: *Gesundheitsberichterstattung des Bundes*, Heft 39, Harninkontinenz, 2007.

Hammes, Christoph, Elmar Heinrich, Tobias Lingenfelder, and Christine Cotic. *Basics Urologie. 4. Auflag*, Elsevier, 2019

Index